A-Level Maths for OCR C4

Paul Sanders

Published in 2005 by:
Nelson Thornes Ltd
Delta Place
27 Bath Road
CHELTENHAM
GL53 7TH
United Kingdom

05 06 07 08 09 / 10 9 8 7 6 5 4 3 2 1

A catalogue record for this book is available from the British Library

ISBN 0 7487 9458 1

Sample paper written by Val Dixon

Page make-up by MCS Publishing Services Ltd, Salisbury, Wiltshire

Printed and bound in Spain by Graphycems

Acknowledgements

We are grateful to the Oxford Cambridge and RSA Examination Board for permission to reproduce all the questions marked OCR.
All answers provided for examination questions are the sole responsibility of the author.

The publishers have made every effort to contact copyright holders but apologise if any have been overlooked.

CONTENTS

INTRODUCTION

A-Level Maths for OCR is a brand new series from Nelson Thornes designed to give you the best chance of success in Advanced Level Maths. This book fully covers the OCR **C4** module specification.

In each chapter, you will find a number of key features:

- A beginning of chapter **OBJECTIVES** section, so you can see clearly what you should learn from each chapter

- **WORKED EXAMPLES** taking you through common questions, step by step

- Carefully graded **EXERCISES** to give you thorough practice in all concepts and skills

- Highlighted **KEY POINTS** to help you see at a glance what you need to know for the exam

- An **IT ICON** **(IT)** to highlight areas where IT software such as Excel may be used

- **EXTENSION** boxes with background information and additional theory

- An end-of-chapter **SUMMARY** to help with your revision

- An end-of-chapter **REVISION EXERCISE** so you can test your understanding of the chapter

At the end of the book, you will find a **MODULE REVISION EXERCISE** containing exam-type questions for the entire module. This is divided into four sections, mirroring the structure of the specification. [Each section tells you which chapters you should have done.]

Finally, there is a **SAMPLE EXAM PAPER** written by an OCR examiner which you can do under timed exam conditions to see just how well prepared you are for the real exam.

1 The Calculus of Trigonometric Functions

The purpose of this chapter is to enable you to

- differentiate trigonometric functions
- use the chain, product and quotient rules to differentiate functions involving trigonometric functions
- integrate simple trigonometric functions

The Derivative of sin x and cos x

So far we have three basic differentiation results:

$$y = x^n \quad \Rightarrow \quad \frac{dy}{dx} = nx^{n-1}$$

$$y = e^{kx} \quad \Rightarrow \quad \frac{dy}{dx} = ke^{kx}$$

$$y = \ln x \quad \Rightarrow \quad \frac{dy}{dx} = \frac{1}{x}.$$

We have also seen how the chain rule, product rule and quotient rule can be used to find the derivatives of more complicated functions that are built up from simple power, exponential or logarithm functions. For example, we can differentiate $\sqrt{4 + e^{2x}}$, $x^2 \ln x$ and $\dfrac{5e^{2x}}{2 + e^{2x}}$.

At the moment we do not know how to find the gradient of any of the trigonometric functions that were introduced in modules C2 and C3.

The diagram below shows the graph of $y = \sin x$, where x is measured in radians.

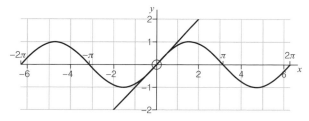

We know that the graph has a maximum point at $\left(\dfrac{\pi}{2}, 1\right)$ so the gradient of the curve $y = \sin x$ at $x = \dfrac{\pi}{2}$ is 0.

Similarly, the graph has a minimum point at $\left(-\dfrac{\pi}{2}, -1\right)$ so the gradient of the curve $y = \sin x$ at $x = -\dfrac{\pi}{2}$ is 0.

The tangent to the curve at the origin has been drawn on the diagram. It suggests that the gradient of the curve $y = \sin x$ at $x = 0$ is 1. This is clearly the greatest gradient at any point on the graph.

Making use of the symmetries of the sine graph, the following table of results can be deduced:

x	-2π	$-\dfrac{3}{2}\pi$	$-\pi$	$-\dfrac{1}{2}\pi$	0	$\dfrac{1}{2}\pi$	π	$\dfrac{3}{2}\pi$	0
Gradient of $y = \sin x$	1	0	−1	0	1	0	−1	0	1

Plotting this information on a graph gives:

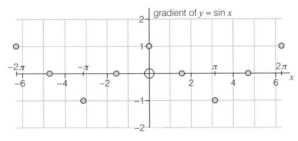

which suggests the graph of the gradient function might be

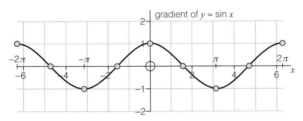

This graph resembles the cosine graph, so we might conjecture that

$$y = \sin x \quad \Longrightarrow \quad \frac{dy}{dx} = \cos x.$$

Looking at the diagrams of $y = \sin x$ and $y = \cos x$, it can be seen that the cosine graph is the image of the sine graph after a translation of $\begin{pmatrix} -\dfrac{\pi}{2} \\ 0 \end{pmatrix}$.

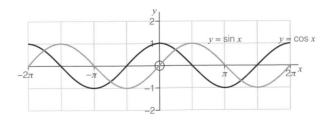

This observation means that the gradients of the cosine graph can be deduced from the gradients of the sine graph to produce the following table of results:

x	-2π	$-\dfrac{3}{2}\pi$	$-\pi$	$-\dfrac{1}{2}\pi$	0	$\dfrac{1}{2}\pi$	π	$\dfrac{3}{2}\pi$	0
Gradient of $y = \cos x$	0	-1	0	1	0	-1	0	1	0

Plotting these points and joining them with a curve gives

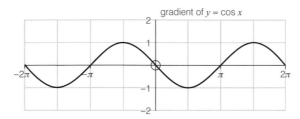

gradient of $y = \cos x$

This graph looks very similar to the graph of $y = -\sin x$ and this leads to the conjecture

$$y = \cos x \quad \Longrightarrow \quad \frac{dy}{dx} = -\sin x.$$

Formal proofs of the results

$$y = \sin x \quad \Longrightarrow \quad \frac{dy}{dx} = \cos x \quad \text{and} \quad y = \cos x \quad \Longrightarrow \quad \frac{dy}{dx} = -\sin x$$

are given in the appendix to this chapter.

It is important to note that these results are **only** valid if **radians** are being used as the unit for angle measurement. Indeed, it is these results that make radians so much more useful than degrees in advanced level mathematics.

The Derivative of sin kx and cos kx

The gradient of $y = \sin 2x$ can be determined using the chain rule:

$$\left. \begin{array}{ccc} x & \rightarrow & 2x & \rightarrow & \sin 2x \\ & & u & \rightarrow & \sin u \end{array} \right\} = y$$

We have $y = \sin u$ where $u = 2x$ so

$$\frac{dy}{dx} = \frac{du}{dx} \times \frac{dy}{du}$$
$$= 2 \times \cos u$$
$$= 2 \cos 2x.$$

$$u = 2x \Longrightarrow \frac{du}{dx} = 2$$
$$y = \sin u \Longrightarrow \frac{dy}{du} = \cos u.$$

The gradient of $y = \cos 5x$ can be determined in a similar way:

$$\left. \begin{array}{ccc} x & \rightarrow & 5x & \rightarrow & \cos 5x \\ & & u & \rightarrow & \cos u \end{array} \right\} = y$$

We have $y = \cos u$ where $u = 5x$ so

$$\frac{dy}{dx} = \frac{du}{dx} \times \frac{dy}{du}$$

$$= 5 \times (-\sin u)$$

$$= -5 \sin 5x.$$

$u = 5x \implies \dfrac{du}{dx} = 5$

$y = \cos u \implies \dfrac{dy}{du} = -\sin u.$

These examples can easily be generalised to give the important results:

> If k is a constant and x is measured in radians then
>
> $$y = \sin kx \implies \frac{dy}{dx} = k \cos kx$$
>
> $$y = \cos kx \implies \frac{dy}{dx} = -k \sin kx$$

EXAMPLE 1

Find the exact value of the gradient of the curve $y = 3 \sin 2x - 5 \cos 2x$ at the point where $x = \dfrac{\pi}{6}$.

SOLUTION

$$y = 3 \sin 2x - 5 \cos 2x$$

$$\implies \frac{dy}{dx} = 3 \times 2 \cos 2x - 5 \times -2 \sin 2x = 6 \cos 2x + 10 \sin 2x.$$

Substituting $x = \dfrac{\pi}{6}$ gives

Recall from C2 that

$$\cos\left(\frac{\pi}{3}\right) = \frac{1}{2}, \ \sin\left(\frac{\pi}{3}\right) = \frac{\sqrt{3}}{2}.$$

$$\frac{dy}{dx} = 6 \cos\left(\frac{\pi}{3}\right) + 10 \sin\left(\frac{\pi}{3}\right) = 6 \times \frac{1}{2} + 10 \times \frac{\sqrt{3}}{2} = 3 + 5\sqrt{3}.$$

The gradient of $y = 3 \sin 2x - 5 \cos 2x$ at $x = \dfrac{\pi}{6}$ is $3 + 5\sqrt{3}$.

EXERCISE 1

1) Find $\dfrac{dy}{dx}$ if

   a) $y = 5 + 4 \cos x$
   b) $y = 5x + 7 \cos 2x$
   c) $y = 5 \sin x + 3 \cos x$
   d) $y = 8 \cos 4x - 3 \sin 4x$
   e) $y = 3x + 4 - 7 \cos 5x$

2) If $x = 4 + 3 \cos 2t$ find the exact value of $\dfrac{dx}{dt}$ when $t = \dfrac{\pi}{3}$.

3) Find the exact value of the gradient of the curve $y = 3 \sin 2x - \cos 4x$ at the point where $x = \dfrac{3\pi}{8}$.

4 Prove that the normal to the curve $y = 2 \cos 3x$ at the point $\left(\dfrac{\pi}{9}, 1\right)$ has equation $27\sqrt{3}y - 9x = 27\sqrt{3} - \pi$.

5 The diagram shows the graphs of $y = x + \sin 2x$ and $y = x$.

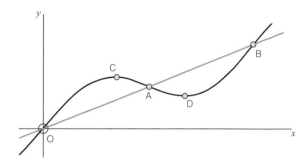

a) Find the co-ordinates of A and B.

b) Find $\dfrac{dy}{dx}$.

c) Find the co-ordinates of the maximum point C and the minimum point D of the curve.

6 Find the x co-ordinates of the stationary points of the curve

$$y = \sqrt{3}x + \cos 2x \qquad 0 \leqslant x \leqslant 2\pi$$

and classify the stationary points as maximum or minimum points.

Further Differentiation of Trigonometric Functions

The product, quotient and chain rules (module C3) can be used to differentiate more complicated functions involving the trigonometric functions.

EXAMPLE 2

Find the derivative of $\quad y = (1 + \sin 4x)^3$.

To work out the value of y we must first work out the value of $1 + \sin 4x$ and then cube the answer. We have the chain of functions:

$$\left. \begin{array}{ccccc} x & \to & 1 + \sin 4x & \to & (1 + \sin 4x)^3 \\ & & u & \to & u^3 \end{array} \right\} = y$$

$$\dfrac{dy}{dx} = \dfrac{du}{dx} \times \dfrac{dy}{du} = 4 \cos 4x \times 3u^2 \qquad\qquad u = 1 + \sin 4x \Rightarrow \dfrac{du}{dx} = 4 \cos 4x$$

$$\Rightarrow \quad \dfrac{dy}{dx} = 12 \cos 4x \times u^2 \qquad\qquad\qquad y = u^3 \Rightarrow \dfrac{dy}{du} = 3u^2.$$

$$\Rightarrow \quad \dfrac{dy}{dx} = 12 \cos 4x (1 + \sin 4x)^2.$$

EXAMPLE 3

Find the derivative of $y = e^{\cos 2x}$.

To work out the value of y we must first work out the value of $\cos 2x$ and then find e to the power of this number. We have the chain of functions:

$$\left.\begin{array}{ccc} x & \to & \cos 2x & \to & e^{\cos 2x} \\ & & u & \to & e^u \end{array}\right\} = y$$

$$\frac{dy}{dx} = \frac{du}{dx} \times \frac{dy}{du} = -2 \sin 2x \times e^u \qquad \boxed{\begin{array}{l} u = \cos 2x \Rightarrow \dfrac{du}{dx} = -2 \sin 2x \\[2mm] y = e^u \Rightarrow \dfrac{dy}{du} = e^u. \end{array}}$$

$$\Rightarrow \quad \frac{dy}{dx} = -2 \sin 2x \, e^{\cos 2x}.$$

EXAMPLE 4

Find the exact value of the gradient of $y = x^2 \sin 3x$ at the point $\left(\dfrac{\pi}{3}, 0\right)$.

$y = u \times v$ where $u = x^2$, $v = \sin 3x$ so the product rule gives

$$\frac{dy}{dx} = \frac{du}{dx} v + u \frac{dv}{dx} \qquad \boxed{\begin{array}{l} u = x^2 \Rightarrow \dfrac{du}{dx} = 2x \\[2mm] v = \sin 3x \Rightarrow \dfrac{dv}{dx} = 3 \cos 3x. \end{array}}$$

$$= 2x \sin 3x + x^2 3 \cos 3x$$

$$= 2x \sin 3x + 3x^2 \cos 3x$$

when $x = \dfrac{\pi}{3}$, $\dfrac{dy}{dx} = 2 \times \dfrac{\pi}{3} \times \sin \pi + 3\left(\dfrac{\pi}{3}\right)^2 \cos \pi = -\dfrac{\pi^2}{3}$.

The Derivative of tan x

Since $\tan x = \dfrac{\sin x}{\cos x}$ the quotient rule can be used to find the gradient of $\tan x$.

If $y = \tan x$ then $y = \dfrac{u}{v}$ where $u = \sin x$, $v = \cos x$ so

$$\frac{dy}{dx} = \frac{\dfrac{du}{dx} v - u \dfrac{dv}{dx}}{v^2} \qquad \boxed{\begin{array}{l} u = \sin x \Rightarrow \dfrac{du}{dx} = \cos x \\[2mm] v = \cos x \Rightarrow \dfrac{dv}{dx} = -\sin x. \end{array}}$$

$$= \frac{\cos x \cos x - \sin x(-\sin x)}{\cos^2 x}$$

$$= \frac{\cos^2 x + \sin^2 x}{\cos^2 x} \qquad \boxed{\text{Remember that } \cos^2 x + \sin^2 x \equiv 1.}$$

$$= \frac{1}{\cos^2 x} \qquad \boxed{\text{Remember that } \dfrac{1}{\cos x} = \sec x.}$$

$$= \sec^2 x.$$

The derivative of $y = \tan kx$, where k is a constant follows from a simple application of the chain rule:

$$\left.\begin{array}{ccccc} x & \longrightarrow & kx & \longrightarrow & \tan kx \\ & & u & \longrightarrow & \tan u \end{array}\right\} = y$$

We have $y = \tan u$ where $u = kx$ so

$$\frac{dy}{dx} = \frac{du}{dx} \times \frac{dy}{du}$$

$$= k \times \sec^2 u$$

$$= k \sec^2 kx.$$

$$u = kx \Rightarrow \frac{du}{dx} = k$$

$$y = \tan u \Rightarrow \frac{dy}{du} = \sec^2 u.$$

If k is a constant and x is measured in radians then

$$y = \tan kx \qquad \Rightarrow \qquad \frac{dy}{dx} = k \sec^2 kx$$

The Derivatives of sec x, cosec x and cot x

Recall from C3 the definitions of sec, cosec and cot:

$$\sec x = \frac{1}{\cos x}, \qquad \operatorname{cosec} x = \frac{1}{\sin x}, \qquad \cot x = \frac{1}{\tan x}.$$

Either the chain rule or the quotient rule can be used to obtain the derivatives of these functions.

If $y = \sec x$ then $y = \dfrac{1}{\cos x} = (\cos x)^{-1}$ so we have the chain

$$\left.\begin{array}{ccccc} x & \longrightarrow & \cos x & \longrightarrow & (\cos x)^{-1} \\ & & u & \longrightarrow & u^{-1} \end{array}\right\} = y$$

We have $y = u^{-1}$ where $u = \cos x$ so

$$\frac{dy}{dx} = \frac{du}{dx} \times \frac{dy}{du}$$

$$= -\sin x \times -1u^{-2}$$

$$= \sin x \times u^{-2}$$

$$= \sin x \times \frac{1}{\cos^2 x}$$

$$= \frac{\sin x}{\cos x} \times \frac{1}{\cos x}$$

$$= \tan x \sec x.$$

$$u = \cos x \Rightarrow \frac{du}{dx} = -\sin x$$

$$y = u^{-1} \Rightarrow \frac{dy}{du} = -1u^{-2}.$$

In a similar way we can also establish the results

$$y = \operatorname{cosec} x \quad \Longrightarrow \quad \frac{dy}{dx} = -\cot x \operatorname{cosec} x$$

and

$$y = \cot x \quad \Longrightarrow \quad \frac{dy}{dx} = -\operatorname{cosec}^2 x.$$

(The proof of these results is an example in the next exercise.)

We now know the derivatives of the six trigonometric functions:

$y = \sin x$	$\Longrightarrow$	$\frac{dy}{dx} = \cos x$	$y = \operatorname{cosec} x$	$\Longrightarrow$	$\frac{dy}{dx} = -\cot x \operatorname{cosec} x$
$y = \cos x$	$\Longrightarrow$	$\frac{dy}{dx} = -\sin x$	$y = \sec x$	$\Longrightarrow$	$\frac{dy}{dx} = \tan x \sec x$
$y = \tan x$	$\Longrightarrow$	$\frac{dy}{dx} = \sec^2 x$	$y = \cot x$	$\Longrightarrow$	$\frac{dy}{dx} = -\operatorname{cosec}^2 x$

EXERCISE 2

1 Find $\frac{dy}{dx}$ if

a) $y = \sin(x^5)$

b) $y = x^4 \sin 2x$

c) $y = \dfrac{\sin 5x}{1 + x^2}$

d) $y = e^{2x} \cos 3x$

e) $y = \sin^5 x = (\sin x)^5$

f) $y = \cos^3(2x)$

g) $y = \dfrac{\sin x}{1 + \cos x}$

h) $y = \dfrac{1}{(3 + \cos 5x)^2}$

2 Find the gradient of the curve $y = (1 + \cos 2x)^3$ at the point $(0, 8)$.

3 A curve has equation $y = (3 + k \sin 2x)^4$ where k is a constant number. The gradient of the curve at the point where $x = 0$ is 54. Find the value of k.

4 Prove that the normal to the curve $y = \dfrac{\sin 2x}{2 + \cos 2x}$ at the point $(\pi, 0)$ has equation $2y + 3x - 3\pi = 0$.

5 Find the gradient of the curve $y = x \sin 2x$ at the point $(\pi, 0)$.

6 a) By writing $\cot x = \dfrac{\cos x}{\sin x}$ and using the quotient rule, prove that

$$\frac{d(\cot x)}{dx} = -\operatorname{cosec}^2 x.$$

b) Use the chain rule to prove that the derivative of $\operatorname{cosec} x$ is $-\cot x \operatorname{cosec} x$.

7 Find the derivatives of

a) $4 \operatorname{cosec} x + 3 \cot x$

b) $5 \sec 2x + \tan 2x$

c) $x^3 \tan 3x$

d) $\sec^5 x$

e) $e^{2x} \sec 5x$

f) $\dfrac{\sec 2x}{1 + \tan 2x}$

8 The diagram shows a sketch of the graph of $y = x \cos 3x$

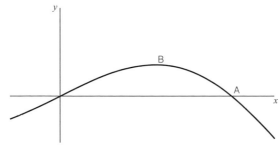

i) Write down the co-ordinates of the point A.

ii) The point B is a stationary point of the curve and has co-ordinates $(\theta, \theta \cos 3\theta)$.

Prove that θ must be a root of the equation $\tan 3x = \dfrac{1}{3x}$.

iii) Prove that the equation $\tan 3x = \dfrac{1}{3x}$ has a root between 0.25 and 0.30.

iv) Show that the equation $\tan 3x = \dfrac{1}{3x}$ can be rewritten as

$$x = \frac{1}{3} \tan^{-1}\left(\frac{1}{3x}\right)$$

and use an iterative process based on this equation, with starting value 0.275, to determine the value of θ correct to three decimal places.

9 If $y = e^{-2t} \cos 3t$ find $\dfrac{d^2y}{dt^2}$ and hence prove that $\dfrac{d^2y}{dt^2} + 4\dfrac{dy}{dt} + 13y = 0$.

10 If $f(x) = x \cos 2x$ find the values of $f'(0)$ and $f''(0)$.

Integration with Trigonometric Functions

We have seen that

$$\frac{d(\sin kx)}{dx} = k \cos kx.$$

Dividing through by k gives

$$\frac{d\left(\frac{1}{k}\sin kx\right)}{dx} = \cos kx.$$

Since integration is simply the reverse of differentiation, this last statement can be written as

$$\int \cos kx \, dx = \frac{1}{k}\sin kx + c.$$

Similarly

$$\frac{d(\cos kx)}{dx} = -k \sin kx$$

and dividing through by $-k$ gives

$$\frac{d\left(-\frac{1}{k} \cos kx\right)}{dx} = \sin kx.$$

This can also be written as an integration statement:

$$\int \sin kx \, dx = -\frac{1}{k} \cos kx + c.$$

Finally, the result

$$\frac{d(\tan kx)}{dx} = k \sec^2 kx$$

$$\implies \quad \frac{d\left(\frac{1}{k} \tan kx\right)}{dx} = \sec^2 kx$$

$$\implies \quad \int \sec^2 kx \, dx = \frac{1}{k} \tan kx + c.$$

We have established three important integration results:

> If k is a constant and x is measured in radians then
>
> $$\int \sin kx \, dx = -\frac{1}{k} \cos kx + c$$
>
> $$\int \cos kx \, dx = \frac{1}{k} \sin kx + c$$
>
> $$\int \sec^2 kx \, dx = \frac{1}{k} \tan kx + c$$

EXAMPLE 5

A curve has gradient $4 \cos 2x + 6 \sec^2 2x$ at the point (x, y) and passes through the point $(0, 5)$. Find the equation of the curve.

SOLUTION

$$\frac{dy}{dx} = 4 \cos 2x + 6 \sec^2 2x$$

$$\implies \quad y = \int (4 \cos 2x + 6 \sec^2 2x) \, dx$$

The integral of $\cos 2x$ is $\frac{1}{2} \cos 2x$.

The integral of $\sec^2 2x$ is $\frac{1}{2} \tan 2x$.

Don't forget the integration constant!

$$\implies \quad y = 4 \times \frac{1}{2} \sin 2x + 6 \times \frac{1}{2} \tan 2x + c$$

$$\implies \quad y = 2 \sin 2x + 3 \tan 2x + c.$$

When $x = 0$, $y = 5$ $\implies$ $5 = 0 + 0 + c$ $\implies$ $c = 5$.

The equation of the curve is $y = 2 \sin 2x + 3 \tan 2x + 5$.

EXAMPLE 6

Find the area of the closed region bounded by the curve $y = 3 + 2\sin 5x$, the y-axis, the x-axis and the line $x = \pi$.

The diagram shows the curve.

It is clear that the value of $3 + 2\sin 5x$ is positive for all values of x, so the area under the curve is given by $\displaystyle\int_0^\pi y\,dx$.

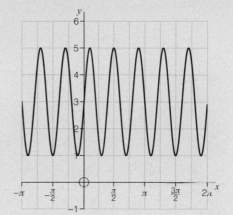

Alternatively, we could argue that since the minimum value of $\sin 5x$ is -1, the minimum value of $3 + 2\sin 5x$ is 1 so the function is certainly positive. The area is therefore found by simply integrating the function.

$$\text{Area} = \int_0^\pi y\,dx = \int_0^\pi (3 + 2\sin 5x)\,dx$$

$$= \left[3x - \frac{2}{5}\cos 5x\right]_0^\pi$$

The integral of $\sin 5x$ is $-\frac{1}{5}\cos 5x$.

$$= \left(3\pi - \frac{2}{5} \times -1\right) - \left(0 - \frac{2}{5}\right)$$

$$= 3\pi + \frac{4}{5}.$$

EXAMPLE 7

Prove that the volume of the solid formed when the curve $y = \tan 2x$ between $x = 0$ and $x = \dfrac{\pi}{6}$ is rotated about the x-axis is $\dfrac{\pi}{6}(3\sqrt{3} - \pi)$.

The volume generated when a curve is rotated about the x-axis is given by $V_x = \displaystyle\int \pi y^2\,dx$.

$$V_y = \pi \int_0^{\frac{\pi}{6}} \tan^2 2x\,dx$$

$$= \pi \int_0^{\frac{\pi}{6}} (\sec^2 2x - 1)\,dx$$

We don't know how to integrate $\tan^2 2x$ but we do know that
$$\sec^2 A \equiv 1 + \tan^2 A$$
or, equivalently
$$\tan^2 A \equiv \sec^2 A - 1.$$

$$= \pi \left[\frac{1}{2}\tan 2x - x\right]_0^{\frac{\pi}{6}}$$

$$= \pi\left(\left(\frac{1}{2}\tan\left(\frac{\pi}{3}\right) - \frac{\pi}{6}\right) - (0 - 0)\right)$$

$$= \pi\left(\frac{1}{2}\sqrt{3} - \frac{\pi}{6}\right)$$

$$= \frac{\pi}{6}(3\sqrt{3} - \pi).$$

EXERCISE 3

1 Find

a) $\displaystyle\int \sin 3x \, dx$

b) $\displaystyle\int 4 \sin 3x + 8 \cos 2x \, dx$

c) $\displaystyle\int \sin 5x - 5 \cos 3x \, dx$

d) $\displaystyle\int 4 \cos 2x - 8 \cos 4x \, dx$

2 If $\dfrac{dx}{dt} = 5 + 3 \cos t$ and $x = 2$ when $t = 0$, find an expression for x in terms of t. Hence find the exact value of x when $t = \dfrac{\pi}{2}$.

3 Evaluate the following definite integrals:

a) $\displaystyle\int_0^{\frac{\pi}{2}} \cos x \, dx$

b) $\displaystyle\int_0^{\pi} \theta + \sin \theta \, d\theta$

c) $\displaystyle\int_0^{\frac{\pi}{2}} \cos 3\theta \, d\theta$

4 Find the area of the closed region bounded by the graph $y = \cos 4x$, the x-axis and the lines $x = 0$ and $x = \dfrac{\pi}{12}$.

5 Find the area of the closed region bounded by the curve $y = \sec^2\left(\dfrac{1}{2}x\right)$, the x-axis, the y-axis and the line $x = \dfrac{1}{2}\pi$.

6 Find the volume of revolution when the portion of the curve $y = \tan x$ between $x = 0$ and $x = \dfrac{1}{4}\pi$ is rotated about the x-axis.

7 Find y if

a) $\dfrac{dy}{dx} = 4 \cos x + 3 \sin x$

b) $\dfrac{dy}{dx} = 9 \cos 3x - 8 \sin 2x$

8 Find $\displaystyle\int 12 \cos 3x + 8 \sin 4x \, dx$

9 Find the volume of revolution when the portion of the curve $y = \sec x$ between $x = 0$ and $x = \dfrac{1}{3}\pi$ is rotated about the x-axis.

10 a) Evaluate $\displaystyle\int_0^{\pi} \cos \theta \, d\theta$.

b) Sketch the graph of $y = \cos \theta$ for $0 \leqslant \theta \leqslant \pi$.
Find the area of the region bounded by the curve $y = \cos \theta$, the θ-axis and the lines $\theta = 0$ and $\theta = \pi$.

11 Sketch the graph of $y = 1 - 2 \sin x$ for $0 \leqslant x \leqslant 2\pi$.
Find the area of the closed region bounded by the graph and the x-axis.

12 On the same set of axes, sketch the graphs of $y = 2 + \cos \theta$ and $y = 1 - \cos \theta$ for values of θ in the domain $0 \leqslant \theta \leqslant 2\pi$.
Find the points of intersection of the two curves.
Find the area between the two curves.

Further Trigonometric Integrations

The identities

$$\cos 2A \equiv 2\cos^2 A - 1 \quad \text{and} \quad \cos 2A \equiv 1 - 2\sin^2 A$$

that were met in module C3 can be rearranged to give

$$\cos^2 A \equiv \frac{1}{2} + \frac{1}{2}\cos 2A \quad \text{and} \quad \sin^2 A \equiv \frac{1}{2} - \frac{1}{2}\cos 2A$$

and these results are frequently useful in integrations.

EXAMPLE 8

Find the volume of revolution when the portion of the curve $y = \sin 3x$ between $x = 0$ and $x = \frac{1}{6}\pi$ is rotated about the x-axis.

S O L U T I O N

$$V_y = \pi \int_0^{\frac{\pi}{6}} \sin^2 3x \, dx$$

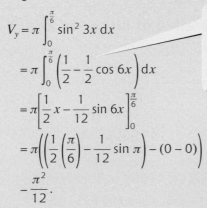

Since $\sin^2 A \equiv \dfrac{1}{2} - \dfrac{1}{2}\cos 2A$

putting $A = 3x$ means we can write

$$\sin^2 3x \equiv \frac{1}{2} - \frac{1}{2}\cos 6x.$$

$$= \pi \int_0^{\frac{\pi}{6}} \left(\frac{1}{2} - \frac{1}{2}\cos 6x \right) dx$$

$$= \pi \left[\frac{1}{2}x - \frac{1}{12}\sin 6x \right]_0^{\frac{\pi}{6}}$$

$$= \pi \left(\left(\frac{1}{2}\left(\frac{\pi}{6} \right) - \frac{1}{12}\sin \pi \right) - (0 - 0) \right)$$

$$\frac{\pi^2}{12}.$$

EXERCISE 4

1 Find

a) $\displaystyle\int \cos^2 x \, dx$

b) $\displaystyle\int \sin^2 x \, dx$

c) $\displaystyle\int \cos^2 4x \, dx$

d) $\displaystyle\int \sin^2 7x \, dx$

e) $\displaystyle\int \sin^2 5x \, dx$

2 Evaluate

i) $\displaystyle\int_0^{\frac{\pi}{8}} \sin^2 2x \, dx$

ii) $\displaystyle\int_0^{\frac{\pi}{12}} \cos^2 3x \, dx$

3 Find the area of the closed region bounded by the curve $y = (1 + \cos x)^2$, the x-axis and the lines $x = \dfrac{\pi}{2}$ and $x = \dfrac{3\pi}{2}$.

4 Find the volume of the solid generated when the curve $y = 2 + \sin 2x$ between $x = 0$ and $x = \pi$ is rotated completely about the x-axis.

5 **a)** By writing $\cos^4 x$ as $\left(\cos^2 x\right)^2$ prove that

$$\cos^4 x \equiv \frac{1}{4}\left(1 + 2\cos 2x + \cos^2 2x\right).$$

b) Deduce that

$$\cos^4 x \equiv \frac{1}{4}\left(\frac{3}{2} + 2\cos 2x + \frac{1}{2}\cos 4x\right).$$

c) Hence evaluate $\displaystyle\int_0^{\frac{\pi}{2}} \cos^4 x \, dx$.

d) Evaluate $\displaystyle\int_0^{\frac{\pi}{2}} \sin^4 x \, dx$.

EXTENSION

Formal Proof of the Derivative of sin x and cos x

The proof of the results will be split into four main steps.

Proposition 1 If θ is measured in radians then $\displaystyle\lim_{\theta \to 0}\left(\frac{\sin\theta}{\theta}\right) = 1$.

Proof

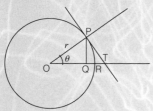

Consider a circle of radius r and a sector OPR of that circle which subtends an acute angle θ radians at O.
Let Q be the foot of the perpendicular from P to OR and let T be the point where the tangent at P meets OR.

It is clear that $PQ = r\sin\theta$ and that $PT = r\tan\theta$.

It is also clear that

$$\text{Area} \triangle OPR < \text{Sector area OPR} < \text{Area} \triangle OPT$$

$$\Rightarrow \quad \frac{1}{2}\times r \times r\sin\theta < \frac{1}{2}r^2\theta < \frac{1}{2}\times r \times r\tan\theta$$

$$\Rightarrow \quad \sin\theta < \theta < \tan\theta.$$

Dividing this inequality through by $\sin\theta$ (which is permissible since θ is acute and $\sin\theta$ is therefore positive), we obtain

$$1 < \frac{\theta}{\sin\theta} < \frac{\tan\theta}{\sin\theta}$$

$$\Rightarrow \quad 1 < \frac{\theta}{\sin\theta} < \frac{1}{\cos\theta}.$$

Inverting this expression gives

$$1 > \frac{\sin\theta}{\theta} > \cos\theta.$$

Note that if a and b are positive numbers with

$$a < b$$

then

$$\frac{1}{a} > \frac{1}{b}.$$

Now consider what happens as $\theta \to 0$:
the left-hand side of the inequality remains fixed at 1;
the right-hand side tends to cos 0 which is 1;
so the middle (which is sandwiched between the left value and the right value) **must also tend to 1**.

We have $\lim\limits_{\theta \to 0} \left(\dfrac{\sin \theta}{\theta} \right) = 1$ as required.

We now find the gradient of the sine curve.

Proposition 2   The gradient of $y = \sin x$ at the point $(p, \sin p)$ is cos p.

Proof

> Recall that the gradient of the curve $y = f(x)$ at the point $(p, f(p))$ is $f'(p)$ where
> $$f'(p) = \lim\limits_{h \to 0} \left(\frac{f(p+h) - f(p)}{h} \right).$$

$$\text{Gradient} = \lim\limits_{0 \to 0} \left(\frac{\sin(p+h) - \sin p}{h} \right)$$

> Using the compound angle formulae (C3)
> $\sin(p+h) \equiv \sin p \cos h + \cos p \sin h$.

$$= \lim\limits_{\theta \to 0} \left(\frac{\sin p \cos h + \cos p \sin h - \sin p}{h} \right)$$

$$= \lim\limits_{\theta \to 0} \left(\frac{\sin p(\cos h - 1) + \cos p \sin h}{h} \right)$$

$$= \lim\limits_{\theta \to 0} \left(\frac{\sin p(\cos h - 1)}{h} \right) + \lim\limits_{h \to 0} \left(\frac{\cos p \sin h}{h} \right)$$

$$= \sin p \lim\limits_{h \to 0} \left(\frac{(\cos h - 1)}{h} \right) + \cos p \lim\limits_{h \to 0} \left(\frac{\sin h}{h} \right). \tag{1}$$

From proposition 1 we know that

$$\lim\limits_{h \to 0} \left(\frac{\sin h}{h} \right) = 1. \tag{2}$$

In module C3, you met the formula

$$\cos 2A = 1 - 2\sin^2 A.$$

Rearranging this gives

$$\cos 2A - 1 = -2\sin^2 A$$

and putting $A = \dfrac{1}{2} h$ now gives

$$\cos h - 1 = -2\sin^2\left(\frac{1}{2} h\right).$$

So

$$\frac{\cos h - 1}{h} = -\frac{2\sin^2\left(\frac{1}{2} h\right)}{h} = -\frac{\sin\left(\frac{1}{2} h\right)}{\frac{1}{2} h}\sin\left(\frac{1}{2} h\right). \tag{3}$$

As $h \to 0$, we know that $\dfrac{\sin\left(\frac{1}{2}h\right)}{\frac{1}{2}h} \to 1$ by proposition 1 and $\sin\left(\frac{1}{2}h\right) \to 0$

so we can deduce from equation [3] that

$$\lim_{h\to0}\left(\frac{\cos h - 1}{h}\right) = 0. \hspace{4cm} [4]$$

Returning to equation [1] and making use of equations [2] and [3], we can now deduce that

$$\text{Gradient} = \sin p \lim_{h\to0}\left(\frac{(\cos h - 1)}{h}\right) + \cos p \lim_{h\to0}\left(\frac{\sin h}{h}\right)$$

$$= \sin p \times 0 + \cos p \times 1$$

$$= \cos p$$

as required.

We now know the gradient of $y = \sin x$ at the point $(p, \sin p)$ is $\cos p$. In other words, we have proved that

$$y = \sin x \implies \frac{\mathrm{d}y}{\mathrm{d}x} = \cos x.$$

Proposition 3 $\sin\left(x + \dfrac{\pi}{2}\right) \equiv \cos x, \qquad \cos\left(x + \dfrac{\pi}{2}\right) \equiv -\sin x.$

These results follow from the compound angle formulae:

$$\sin\left(x + \frac{\pi}{2}\right) \equiv \sin x \cos\left(\frac{\pi}{2}\right) + \cos x \sin\left(\frac{\pi}{2}\right)$$

$$\equiv \sin x \times 0 + \cos x \times 1$$

$$\equiv \cos x$$

$$\cos\left(x + \frac{\pi}{2}\right) \equiv \cos x \cos\left(\frac{\pi}{2}\right) - \sin x \sin\left(\frac{\pi}{2}\right)$$

$$\equiv \cos x \times 0 - \sin x \times 1$$

$$\equiv -\sin x.$$

Proposition 4   If $y = \cos x$ then $\dfrac{\mathrm{d}y}{\mathrm{d}x} = -\sin x.$

We know from proposition 3 that $\cos x \equiv \sin\left(x + \dfrac{\pi}{2}\right).$

$y = \cos x = \sin\left(x + \dfrac{\pi}{2}\right)$ can be differentiated using the chain rule and the result of proposition 2:

$$\left.\begin{array}{ccc} x & \to & x + \dfrac{\pi}{2} & \to & \sin\left(x + \dfrac{\pi}{2}\right) \\[2mm] & & u & & \sin u \end{array}\right\} = y$$

We have $y = \sin u$ where $u = x + \dfrac{\pi}{2}$ so

$$\dfrac{dy}{dx} = \dfrac{du}{dx} \times \dfrac{dy}{du} = 1 \times \cos u$$

$$\implies \dfrac{dy}{dx} = \cos u$$

$$\implies \dfrac{dy}{dx} = \cos\left(x + \dfrac{\pi}{2}\right)$$

$$\implies \dfrac{dy}{dx} = -\sin x \qquad \text{(using the second half of proposition 3).}$$

Having studied this chapter you should know how to

- differentiate trigonometric functions using the results

$y = \sin x$	$\implies$	$\dfrac{dy}{dx} = \cos x$	$y = \sin kx \implies \dfrac{dy}{dx} = k\cos kx$	
$y = \cos x$	$\implies$	$\dfrac{dy}{dx} = -\sin x$	$y = \cos kx \implies \dfrac{dy}{dx} = -k\sin kx$	
$y = \tan x$	$\implies$	$\dfrac{dy}{dx} = \sec^2 x$	$y = \tan kx \implies \dfrac{dy}{dx} = k\sec^2 kx$	
$y = \operatorname{cosec} x$	$\implies$	$\dfrac{dy}{dx} = -\cot x \operatorname{cosec} x$		
$y = \sec x$	$\implies$	$\dfrac{dy}{dx} = \tan x \sec x$		
$y = \cot x$	$\implies$	$\dfrac{dy}{dx} = -\operatorname{cosec}^2 x$		

- integrate trigonometric functions using the results

$$\int \sin kx \, dx = -\frac{1}{k}\cos kx + c$$

$$\int \cos kx \, dx = \frac{1}{k}\sin kx + c$$

$$\int \sec^2 kx \, dx = \frac{1}{k}\tan kx + c$$

- use the identities $\cos^2 A \equiv \dfrac{1}{2} + \dfrac{1}{2}\cos 2A$ and $\sin^2 A \equiv \dfrac{1}{2} - \dfrac{1}{2}\cos 2A$ to integrate $\sin^2 kx$ and $\cos^2 kx$

REVISION EXERCISE

1 If $y = 5 \cos 2x + 3 \sin 2x$ find $\dfrac{\mathrm{d}y}{\mathrm{d}x}$ and $\dfrac{\mathrm{d}^2 y}{\mathrm{d}x^2}$.

Show that $\dfrac{\mathrm{d}^2 y}{\mathrm{d}x^2} + 4y = 0$.

2 Differentiate $\dfrac{\tan x}{x}$ with respect to x.

(OCR Jun 1999 P3)

3 **a)** Differentiate with respect to x

$x^2 \cos 4x$.

b) Find the exact value of $\displaystyle\int_0^{\frac{1}{4}\pi} 5 \cos 3x + 4 \sin 2x \,\mathrm{d}x$.

(OCR Mar 1999 P3)

4 Find the exact value of the gradient of the curve $y = \dfrac{4\sqrt{3} \sin x}{1 + 2 \cos x}$ at the point where $x = \dfrac{1}{3}\pi$.

5 **i)** Find the value of $\displaystyle\int_0^{\frac{1}{4}\pi} \sec^2 x \,\mathrm{d}x$.

ii) A region R in the first quadrant is bounded by the curve $y = \tan x$, the x-axis and the line $x = \dfrac{1}{4}\pi$. Show that the exact value of the volume of the solid formed when R is rotated completely about the x-axis is $\pi - \dfrac{1}{4}\pi^2$.

(OCR Nov 1998 P3)

6 The region R, shown shaded in the diagram, is bounded by the curve $y = 4 \sin 2x$, the x-axis and the line $x = \dfrac{5}{12}\pi$.

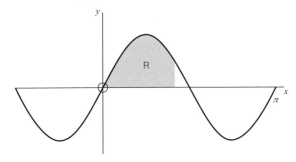

a) Find the exact value of the area of the region R.
b) Find the exact value of the volume of the solid formed when R is rotated completely about the x-axis.

7   If $y = e^{-3x} \sin 5x$ find the value of $\dfrac{d^2y}{dx^2}$ when $x = 0$.

8   The diagram shows the graph of $y = e^{-x} \sin x \qquad 0 \leqslant x \leqslant 2\pi$.

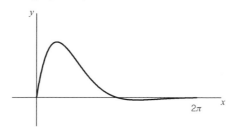

Find the co-ordinates of the stationary points of the curve.

9   Find the equation of the tangent to the curve $y = \sec 2x$ at the point where $x = \dfrac{1}{6}\pi$.

10   A curve C is such that its gradient at the point (x, y) is given by

$$\frac{dy}{dx} = 6 \cos 2x + 8 \sin 2x$$

and the curve passes through the point $(\pi, 5)$.
Find the co-ordinates of the point where the curve meets the y-axis.

11   It is given that

$$f(x) = \cos 6x \quad \text{and} \quad g(x) = 2 \sin^2 3x.$$

   a)   Write down $f'(x)$.
   b)   Find $g'(x)$.
   c)   Show that $f'(x) + g'(x) = 0$.
   d)   Explain this result by reference to $f(x)$ and $g(x)$.

12   The equation of a curve C is $y = \dfrac{x}{\sin 2x}$.

   a)   Prove that the curve has gradient -1 when $x = \dfrac{3}{4}\pi$.

   b)   Find the equation of the normal to the curve C at the point where $x = \dfrac{3}{4}\pi$.

13   Find $\dfrac{dy}{dt}$ if

   a) $y = t^2 \tan 3t$ **b)** $y = \ln(2 + \cos 3t)$

14 Differentiate $y = \dfrac{1}{x}\sec x$.

The diagram shows a graph of $y = \dfrac{1}{x}\sec x$ for $0 < x < \dfrac{\pi}{2}$.

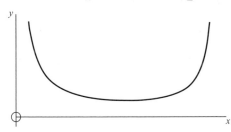

The point $\left(\theta, \dfrac{1}{\theta}\sec\theta\right)$ is a stationary point on the curve $y = \dfrac{1}{x}\sec x$ with $0 < \theta < \dfrac{1}{2}\pi$.

Prove that θ must satisfy the equation $\tan\theta - \dfrac{1}{\theta} = 0$.

Show that this equation can be rewritten as $\theta = \tan^{-1}\left(\dfrac{1}{\theta}\right)$ and use an iterative process

based on this equation, with starting value 1, to determine the value of θ correct to three decimal places.

2 Techniques of Integration 1

The purpose of this chapter is to enable you to

- integrate expressions of the form $\dfrac{f'(x)}{f(x)}$

- use partial fractions to integrate expressions of the form
  $$\frac{px+q}{(x+\alpha)(x+\beta)}$$

- use integration by substitution to determine a variety of integrals

Integrals of Expressions of the Form $\dfrac{f'(x)}{f(x)}$

In C3 the results

$$\frac{d}{dx}(\ln x) = \frac{1}{x} \quad \text{and} \quad \frac{d}{dx}(\ln|x|) = \frac{1}{x}$$

were met.

The chain rule together with these two results allows any expression of the form $\ln(f(x))$ or $\ln(|f(x)|)$ to be differentiated.

EXAMPLE 1

Find $\dfrac{dy}{dx}$ if $y = \ln(3 + \sin 2x)$.

SOLUTION

The evaluation of y from a given value of x can be illustrated by the chain of functions

$$\left. \begin{array}{ccccc} x & \to & 3 + \sin 2x & \to & \ln(3 + \sin 2x) \\ & & u & \to & \ln u \end{array} \right\} = y$$

So $y = \ln u$ where $u = 3 + \sin 2x$. The chain rule now gives

$$\frac{dy}{dx} = \frac{du}{dx} \times \frac{dy}{du} = 2\cos 2x \times \frac{1}{u}$$

$$\Rightarrow \quad \frac{dy}{dx} = \frac{2\cos 2x}{u}$$

$$\Rightarrow \quad \frac{dy}{dx} = \frac{2\cos 2x}{3 + \sin 2x}.$$

$$u = 3 + \sin 2x \Rightarrow \frac{du}{dx} = 2\cos 2x$$

$$y = \ln u \Rightarrow \frac{dy}{du} = \frac{1}{u}.$$

EXAMPLE 2

Find $\dfrac{dy}{dx}$ if $y = \ln(|5 - x^2|)$.

The evaluation of y from a given value of x can be illustrated by the chain of functions

$$\left.\begin{array}{ccc} x & \rightarrow & 5 - x^2 & \rightarrow & \ln(|5 - x^2|) \\ & & u & & \ln|u| \end{array}\right\} = y$$

So $y = \ln|u|$ where $u = 5 - x^2$. The chain rule now gives

$$\frac{dy}{dx} = \frac{du}{dx} \times \frac{dy}{du} = -2x \times \frac{1}{u}$$

$$\Rightarrow \quad \frac{dy}{dx} = \frac{-2x}{u}$$

$$\Rightarrow \quad \frac{dy}{dx} = \frac{-2x}{5 - x^2}.$$

$$u = 5 - x^2 \Rightarrow \frac{du}{dx} = -2x$$

$$y = \ln|u| \Rightarrow \frac{dy}{du} = \frac{1}{u}.$$

EXERCISE 1

Find the derivatives of the following functions:

1. $\ln(x^4 + 1)$

2. $\ln(|5x - 3|)$

3. $\ln(x^2 + 4)$

4. $\ln(5 + 3\sin x)$

5. $\ln(2 + \cos 4x)$

6. $\ln(|e^{3x} - 5|)$

The examples of the previous section and exercise 1 can readily be generalised.

If $y = \ln(|f(x)|)$ then $y = \ln|u|$ where $u = f(x)$ and the chain rule gives

$$\frac{dy}{dx} = \frac{du}{dx} \times \frac{dy}{du} = f'(x) \times \frac{1}{u}$$

$$\Rightarrow \quad \frac{dy}{dx} = \frac{f'(x)}{u}$$

$$\Rightarrow \quad \frac{dy}{dx} = \frac{f'(x)}{f(x)}.$$

$$u = f(x) \Rightarrow \frac{du}{dx} = f'(x)$$

$$y = \ln|u| \Rightarrow \frac{dy}{du} = \frac{1}{u}.$$

We have proved that $\quad y = \ln(|f(x)|) \quad \Rightarrow \quad \dfrac{dy}{dx} = \dfrac{f'(x)}{f(x)}.$

Since integration is the reverse of differentiation, the following two results may be deduced.

For any function f $\quad \displaystyle\int \frac{f'(x)}{f(x)}\,dx = \ln(|f(x)|) + c.$

If the function f is **known to take only positive values** then $\displaystyle\int \frac{f'(x)}{f(x)}\,dx = \ln(f(x)) + c.$

We have established a result that will prove to be very useful in the future:

$$\int \frac{f'(x)}{f(x)}\,dx = \ln(|f(x)|) + c$$

> You can omit the modulus signs if you are sure that $f(x)$ is positive.

To use this result directly, the expression to be integrated needs to be a fraction where the top line is the derivative of the bottom line.

EXAMPLE 3

Find $\displaystyle\int \frac{2\cos x}{5 + 2\sin x}\,dx.$

In this case the top of the fraction is the derivative of the bottom of the fraction so we can immediately write

$$\int \frac{2\cos x}{5 + 2\sin x}\,dx = \ln(5 + 2\sin x) + c.$$

> The smallest possible value of $\sin x$ is -1 so the smallest possible value of $5 + 2\sin x$ is 3.
>
> $5 + 2\sin x$ is therefore always positive and the modulus signs can be omitted in the final result.

EXAMPLE 4

Find $\displaystyle\int \frac{4e^{2x}}{2e^{2x} - 1}\,dx.$

In this case the top of the fraction is the derivative of the bottom so we can immediately write

$$\int \frac{4e^{2x}}{2e^{2x} - 1}\,dx = \ln(|2e^{2x} - 1|) + c.$$

> $2e^{2x} - 1$ is not always positive so the modulus signs must be included in the final result.

EXAMPLE 5

Prove that $\displaystyle\int_0^1 \frac{4}{4x - 6}\,dx = -\ln 3.$

$$\int_0^1 \frac{4}{4x - 6}\,dx = \left[\ln(|4x - 6|)\right]_0^1$$

> Between 0 and 1, $4x - 6$ is negative so the modulus signs are essential.

$$= \ln|-2| - \ln|-6|$$
$$= \ln 2 - \ln 6$$
$$= \ln\left(\frac{1}{3}\right)$$
$$= \ln(3^{-1})$$
$$= -\ln 3.$$

> Remember the properties of logarithms:
>
> $\ln x - \ln y = \ln\left(\dfrac{x}{y}\right)$
>
> $\ln(x^p) = p\ln x.$

This integration result

$$\int \frac{f'(x)}{f(x)}\,dx = \ln(|f(x)|) + c$$

can also be used when the expression to be integrated is a fraction where the top line is a multiple of the derivative of the bottom line.

EXAMPLE 6

Find $\displaystyle\int \frac{8x}{x^2 + 1}\,dx.$

In this case the top of the fraction is four times the derivative of the bottom so we can immediately write

$$\int \frac{8x}{x^2 + 1}\,dx = \int 4 \times \frac{2x}{x^2 + 1}\,dx = 4\ln(x^2 + 1) + c.$$

$x^2 + 1$ is always positive so the modulus signs are not essential.

EXAMPLE 7

Find $\displaystyle\int \frac{x^2 + 2}{x^3 + 6x + 5}\,dx.$

The derivative of the bottom line is $3x^2 + 6$ so the top of the fraction is one-third of the derivative of the bottom:

$$\int \frac{x^2 + 2}{x^3 + 6x + 5}\,dx = \int \frac{1}{3} \times \frac{3x^2 + 6}{x^3 + 6x + 5}\,dx = \frac{1}{3}\ln(|x^3 + 6x + 5|) + c.$$

$x^3 + 6x + 5$ is not always positive so the modulus signs must be included in the final result.

In the previous chapter the integrals of sin kx and cos kx were introduced.

We can now find the integral of tan kx.

EXAMPLE 8

Find $\displaystyle\int \tan 3x\,dx.$

$$\int \tan 3x\,dx = \int \frac{\sin 3x}{\cos 3x}\,dx.$$

The derivative of cos $3x$ is $-3\sin 3x$ so the top of the fraction is $(-1/3)$ of the derivative of the bottom of the fraction:

$$\int \tan 3x\,dx = \int \frac{\sin 3x}{\cos 3x}\,dx = \int -\frac{1}{3} \times \frac{-3\sin 3x}{\cos 3x}\,dx = -\frac{1}{3}\ln(|\cos 3x|) + c.$$

The argument of this example can be generalised to give the standard integral:

$$\int \tan kx \, dx = -\frac{1}{k} \ln |\cos kx| + c = \frac{1}{k} \ln |\sec kx| + c$$

Note that
$$-\frac{1}{k} \ln (|\cos kx|) = \frac{1}{k} \ln (|\cos kx|^{-1})$$
$$= \frac{1}{k} \ln (|\sec kx|)$$

EXERCISE 2

Find the following integrals:

1 $\displaystyle\int \frac{2x}{x^2 + 4} \, dx$

2 $\displaystyle\int \frac{4}{4x - 3} \, dx$

3 $\displaystyle\int \frac{3x^2}{x^3 + 16} \, dx$

4 $\displaystyle\int \frac{2 \sin 2x}{3 - \cos 2x} \, dx$

5 $\displaystyle\int \frac{6e^x}{1 + e^x} \, dx$

6 $\displaystyle\int \frac{6 \cos x}{5 + 2 \sin x} \, dx$

7 $\displaystyle\int \frac{10 \sin 2x}{3 + \cos 2x} \, dx$

8 $\displaystyle\int \frac{12e^{2x}}{3e^{2x} + 4} \, dx$

9 $\displaystyle\int \frac{x}{x^2 + 16} \, dx$

10 $\displaystyle\int \frac{x + 1}{x^2 + 2x + 4} \, dx$

11 $\displaystyle\int \cot 5x \, dx$

12 $\displaystyle\int \frac{e^{3x}}{1 + e^{3x}} \, dx$

13 $\displaystyle\int \frac{e^{-3x}}{4 + e^{-3x}} \, dx$

14 $\displaystyle\int_0^2 \frac{x}{x^2 + 4} \, dx$

15 $\displaystyle\int_0^{\frac{\pi}{4}} \tan \theta \, d\theta$

16 $\displaystyle\int_0^1 \frac{1}{3 + 2t} \, dt$

17 $\displaystyle\int_0^2 \frac{3}{u - 4} \, du$

18 $\displaystyle\int_3^4 \frac{x}{4 - x^2} \, dx$

19 Sketch the graph of $y = \dfrac{1}{2x - 1}$ taking care to show all the important features of the graph.

Find the area of the closed region bounded by the x-axis, the curve and the lines $x = 1$ and $x = 3$.

20 a) Sketch the graph of $y = \dfrac{3x - 2}{x - 3}$ taking care to show all the important features of the graph.

b) Find values P and Q so that $\dfrac{3x - 2}{x - 3} = P + \dfrac{Q}{x - 3}$.

c) Find the area of the closed region bounded by the x-axis, the curve and the lines $x = 4$ and $x = 6$.

21 a) Find the stationary points of the curve $y = x + \dfrac{4}{x - 2}$ and hence sketch the curve.

b) Find the area of the closed region formed by the line $y = 7$ and the curve.

22 Write $\dfrac{5}{x + 1} + \dfrac{2}{x}$ as a single fraction.

Hence find the value of $\displaystyle\int_1^3 \frac{7x + 2}{x(x + 1)} \, dx$.

23 Write $\dfrac{4}{2x + 1} - \dfrac{1}{x - 3}$ as a single fraction.

Hence find the value of $\displaystyle\int \frac{2x - 13}{(2x + 1)(x - 3)} \, dx$.

A First Look at Partial Fractions

The fractions $\dfrac{3}{x+2}$ and $\dfrac{5}{x-3}$ may be added together using $(x+2)(x-3)$ as a common denominator:

$$\frac{3}{x+2}+\frac{5}{x-3}=\frac{3(x-3)}{(x+2)(x-3)}+\frac{5(x+2)}{(x+2)(x-3)}=\frac{3x-9+5x+10}{(x+2)(x-3)}=\frac{8x+1}{(x+2)(x-3)}.$$

The process can also be reversed.

For example, it is reasonable to expect that the single fraction $\dfrac{5x+13}{(x+1)(x+3)}$ can be written as the sum of two fractions: one of which has a denominator of $(x+1)$ and the other having a denominator of $(x+3)$.

We want to find values P and Q so that

$$\frac{5x+13}{(x+1)(x+3)} \equiv \frac{P}{x+1}+\frac{Q}{x+3}. \qquad [1]$$

Since

$$\frac{P}{(x+1)}+\frac{Q}{(x+3)} \equiv \frac{P(x+3)}{(x+1)(x+3)}+\frac{Q(x+1)}{(x+1)(x+3)} \equiv \frac{P(x+3)+Q(x+1)}{(x+1)(x+3)},$$

values P and Q must be found so that

$$\frac{5x+13}{(x+1)(x+3)} \equiv \frac{P(x+3)+Q(x+1)}{(x+1)(x+3)}.$$

Since the two identical fractions have the same denominator, their numerators must be identical:

$$\Rightarrow \quad 5x+13 \equiv P(x+3)+Q(x+1). \qquad [2]$$

> In practice, we will usually move from identity [1] through to identity [2] by multiplying equation [1] by the common denominator of the fraction to be expanded.

This identity must hold for **all values of x**.
In particular the identity must be valid when $x=-3$ and when $x=-1$.

Putting $x=-3$ into [2] gives

$$-2 = P \times 0 + Q \times -2 \Rightarrow Q = 1.$$

> These values are chosen since they enable the values of P and Q to be determined immediately from equation [2].

Putting $x=-1$ into [2] gives

$$8 = P \times 2 + Q \times 0 \Rightarrow P = 4.$$

The values of P and Q in identity [1] have now been found, so $\dfrac{5x+13}{(x+1)(x+3)} \equiv \dfrac{4}{x+1}+\dfrac{1}{x+3}.$

When we write $\dfrac{Ax+B}{(x+a)(x+b)}$ in its equivalent form $\dfrac{P}{x+a}+\dfrac{Q}{x+b}$ we say that we have

expressed $\dfrac{Ax+B}{(x+a)(x+b)}$ as **a sum of two partial fractions**.

One immediate application of this result is integration: we can now write

$$\int \frac{5x + 13}{(x+1)(x+3)} \, dx = \int \left(\frac{4}{x+1} + \frac{1}{x+3} \right) dx$$

$$= \int \left(4 \times \frac{1}{x+1} + \frac{1}{x+3} \right) dx$$

$$= 4 \ln(|x+1|) + \ln(|x+3|) + c.$$

Using the properties of logarithms, it is possible to write

$$4 \ln(|x+1|) + \ln(|x+3|) + c = \ln(|x+1|^4) + \ln(|x+3|) + c$$

$$= \ln(|x+1|^4 | x+3|) + c$$

so the final answer for the integration could be rewritten as

$$\int \frac{5x + 13}{(x+1)(x+3)} \, dx = \ln(|x+1|^4 |x+3|) + c.$$

EXAMPLE 9

Express $\dfrac{7x + 6}{(x-2)(x+3)}$ as a sum of partial fractions and hence prove that

$$\int_3^4 \frac{7x + 6}{(x-2)(x+3)} \, dx = \ln\left(\frac{686}{27} \right).$$

SOLUTION

We want to find values P and Q so that $\dfrac{7x + 6}{(x-2)(x+3)} \equiv \dfrac{P}{x-2} + \dfrac{Q}{x+3}$.

Multiplying through by the common denominator, $(x-2)(x+3)$,

$$\Rightarrow \quad 7x + 6 \equiv P(x+3) + Q(x-2).$$

We want this to be valid **for all values of x**. In particular it must be true when $x = 2$ and when $x = -3$.

Putting $x = 2$ gives

$$20 = 5P \Rightarrow P = 4.$$

Putting $x = -3$ gives

$$-15 = -5Q \Rightarrow Q = 3.$$

So

$$\frac{7x + 6}{(x-2)(x+3)} \equiv \frac{4}{x-2} + \frac{3}{x+3}.$$

EXAMPLE 9 (continued)

The integration can now be calculated using the partial fraction form of the integrand:

$$\int_3^4 \frac{7x+6}{(x-2)(x+3)}\,dx = \int_3^4 \frac{4}{x-2} + \frac{3}{x+3}\,dx$$

$$= \left[4\ln(x-2) + 3\ln(x+3)\right]_3^4$$

$$= (4\ln 2 + 3\ln 7) - (4\ln 1 + 3\ln 6)$$

$$= 4\ln 2 + 3\ln 7 - 3\ln 6$$

$$= \ln(2^4) + \ln(7^3) - \ln(6^3)$$

$$= \ln\left(\frac{2^4 \times 7^3}{6^3}\right)$$

$$= \ln\left(\frac{5488}{216}\right)$$

$$= \ln\left(\frac{686}{27}\right).$$

EXERCISE 3

1 Express the following as a sum of two partial fractions:

a) $\dfrac{3x+7}{(x+3)(x+5)}$ **b)** $\dfrac{4x}{x^2-9}$ **c)** $\dfrac{7x+9}{3x^2+2x-5}$

2 Find $\displaystyle\int \frac{x-13}{(x-3)(x+2)}\,dx.$

3 Prove that $\displaystyle\int_0^1 \frac{1}{(x+1)(2x+1)}\,dx = \ln\left(\tfrac{3}{2}\right).$

4 If $f(x) = \dfrac{8x-26}{x^2-4x+3}$

a) express $f(x)$ as a sum of partial fractions;

b) hence, or otherwise, find $f'(x)$ and prove that

$$f''(x) = \frac{18}{(x-1)^3} - \frac{2}{(x-3)^3};$$

c) find the turning points of the graph of $f(x)$ and sketch the graph;

d) find the area of the region bounded by the x-axis, the graph of $f(x)$ and the lines $x = 5$ and $x = 7$.

5 The diagram shows the graph of

$$y = \frac{x+7}{(x+1)(x-2)}.$$

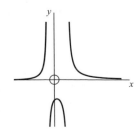

a) Write down the equations of the asymptotes of this graph.

b) Prove that the closed region bounded by the graph, the co-ordinate axes and the line $x = 1$ has an area of $5\ln 2$ units^2.

Integration by Substitution

In module C3, simple cases of integration by substitution were considered. In this section the technique will be applied to a larger variety of integrals.

The first two examples are similar to examples considered in C3.

EXAMPLE 10

Find $\displaystyle\int \frac{1}{(3x+1)^4}\,dx$.

$$\int \frac{1}{(3x+1)^4}\,dx = \int \frac{1}{u^4}\frac{1}{3}\,du$$

$$= \int \frac{1}{3}u^{-4}\,du$$

$$= \frac{1}{3}\times\frac{1}{-3}u^{-3} + c$$

$$= -\frac{1}{9u^3} + c$$

$$= -\frac{1}{9(3x+1)^3} + c.$$

If the substitution $u = 3x + 1$ is used then

$\dfrac{1}{(3x+1)^4}$ becomes $\dfrac{1}{u^4}$ or u^{-4}

and $\dfrac{du}{dx} = 3 \Rightarrow \text{`}dx = \dfrac{1}{3}du\text{'}$.

EXAMPLE 11

Evaluate $\displaystyle\int_0^1 \frac{1}{\sqrt{5x+4}}\,dx$.

$$\int_0^1 \frac{1}{\sqrt{5x+4}}\,dx = \int_4^9 \frac{1}{\sqrt{u}}\frac{1}{5}\,du$$

$$= \int_4^9 \frac{1}{5}u^{-\frac{1}{2}}\,du$$

$$= \left[\frac{1}{5}\times\frac{1}{\frac{1}{2}}u^{1/2}\right]_4^9$$

$$= \left[\frac{2}{5}u^{1/2}\right]_4^9$$

$$= \frac{2}{5}\times 3 - \frac{2}{5}\times 2$$

$$= \frac{2}{5}.$$

If the substitution $u = 5x + 4$ is used then

$\dfrac{1}{\sqrt{5x+4}}$ becomes $\dfrac{1}{\sqrt{u}}$ or $u^{-\frac{1}{2}}$

and $\dfrac{du}{dx} = 5 \Rightarrow \text{`}dx = \dfrac{1}{5}du\text{'}$.

Considering the limits, since $u = 5x + 4$

when $x = 0$, $u = 5 \times 0 + 4 = 4$

and

when $x = 1$, $u = 5 \times 1 + 4 = 9$.

You may be able to use your calculator to check the values of definite integrals.

Both the substitutions used so far have been linear substitutions of the form $u = ax + b$. Other substitutions can be used.

EXAMPLE 12

SOLUTION

Find $\displaystyle\int_0^4 x\sqrt{x^2+9}\;dx$.

$$\int_0^4 x\sqrt{x^2+9}\;dx = \int_9^{25} x\sqrt{u}\,\frac{1}{2x}\,du$$

$$= \int_9^{25} \sqrt{u}\,\frac{1}{2}\,du$$

$$= \int_9^{25} \frac{1}{2}\,u^{\frac{1}{2}}\,du$$

$$= \left[\frac{1}{2}\times\frac{1}{\frac{3}{2}}\,u^{\frac{3}{2}}\right]_9^{25}$$

$$= \left[\frac{1}{3}\,u^{\frac{3}{2}}\right]_9^{25}$$

$$= \frac{125}{3} - \frac{27}{3}$$

$$= \frac{98}{3}.$$

If the substitution $u = x^2 + 9$ is made then

$$\sqrt{x^2+9} = \sqrt{u} = u^{\frac{1}{2}}$$

and

$$\frac{du}{dx} = 2x \Rightarrow \text{`}dx = \frac{1}{2x}\,du\text{'}.$$

When $x = 0$, $u = 0^2 + 9 = 9$
and when $x = 4$, $u = 4^2 + 9 = 25$.

EXAMPLE 13

SOLUTION

Evaluate $\displaystyle\int \frac{x^3}{(2x^2+1)^2}\;dx$.

$$\int \frac{x^3}{(2x^2+1)^2}\;dx = \int \frac{x^3}{u^2}\,\frac{1}{4x}\,du$$

$$= \int \frac{1}{4}\times x^2\times\frac{1}{u^2}\,du.$$

$$= \int \frac{1}{4}\times\frac{1}{2}\,(u-1)\times\frac{1}{u^2}\,du$$

$$= \int \frac{1}{8}\times\frac{u-1}{u^2}\,du$$

$$= \frac{1}{8}\int \left(\frac{1}{u}-\frac{1}{u^2}\right)\,du$$

$$= \frac{1}{8}\left(\ln(|\,u\,|)+\frac{1}{u}\right)+c$$

$$= \frac{1}{8}\ln(|\,2x^2+1\,|)+\frac{1}{8(2x^2+1)}+c.$$

If the substitution $u = 2x^2 + 1$ is made then

$$\frac{du}{dx} = 4x \Rightarrow \text{`}dx = \frac{1}{4x}\,du\text{'}.$$

We have still got two variables in the integral.

x can be removed from the integral by rearranging the substitution formula:

$$u = 2x^2 + 1$$
$$\Rightarrow \quad 2x^2 = u - 1$$
$$\Rightarrow \quad x^2 = \frac{1}{2}(u-1).$$

The modulus signs could be omitted here since $2x^2 + 1$ is always positive.

EXERCISE 4

Use an appropriate substitution to find the following integrals:

1 $\displaystyle\int (2x-5)^6\,dx$

2 $\displaystyle\int (4x+1)^3\,dx$

3 $\displaystyle\int \sqrt{6x+5}\,dx$

4 $\displaystyle\int x^2(2x^3+1)^4\,dx$

5 $\displaystyle\int \cos x(1+\sin x)^5\,dx$

6 $\displaystyle\int \sin 2x\,(3+\cos 2x)^3\,dx$

7 $\displaystyle\int e^{2x}\sqrt{4+e^{2x}}\,dx$

8 $\displaystyle\int_1^3 \frac{x-1}{(x+1)^3}\,dx$

9 $\displaystyle\int_0^4 \frac{x}{\sqrt{2x+1}}\,dx$

10 $\displaystyle\int_0^2 \frac{x}{\sqrt{2x^2+1}}\,dx$

11 $\displaystyle\int_0^{\frac{\pi}{4}} \cos 2x\,\sin^4 2x\,dx$

12 $\displaystyle\int_0^{\ln 2} \frac{e^{3x}}{(1+e^{3x})^2}\,dx$

13 $\displaystyle\int_0^{16} \frac{x}{\sqrt{x+9}}\,dx$

14 $\displaystyle\int_0^{\frac{\pi}{4}} (1+2\tan x)^3 \sec^2 x\,dx$

15 $\displaystyle\int_0^3 \frac{x^3}{(x^2+16)^2}\,dx$

EXTENSION

The Integrals of tan x, cot x, sec x and cosec x

We have already seen that

$$\int \tan x\,dx = \int \frac{\sin x}{\cos x}\,dx$$

$$= -\int \frac{-\sin x}{\cos x}\,dx$$

$$= -\ln|\cos x| + c = \ln|\sec x| + c.$$

Similarly, since $\cot x = \dfrac{\cos x}{\sin x}$, we can write

$$\int \cot x\,dx = \int \frac{\cos x}{\sin x}\,dx$$

$$= \ln|\sin x| + c.$$

Since the derivative of sec x is tan x sec x and the derivative of tan x is $\sec^2 x$, we know that if

$$f(x) = \sec x + \tan x$$

then

$$f'(x) = \tan x \sec x + \sec^2 x$$

$$= \sec x(\tan x + \sec x)$$

$$= \sec x\, f(x)$$

and this can be rewritten as

$$\sec x = \frac{f'(x)}{f(x)}.$$

We can now write

$$\int \sec x \, dx = \int \frac{f'(x)}{f(x)} \, dx$$
$$= \ln|f(x)| + c$$
$$= \ln|\sec x + \tan x| + c.$$

Finally, consider the function g defined by

$$g(x) = \cot x + \operatorname{cosec} x.$$

Since the derivative of $\cot x$ is $-\operatorname{cosec}^2 x$ and the derivative of $\operatorname{cosec} x$ is $-\cot x \operatorname{cosec} x$, we can write

$$g'(x) = -\operatorname{cosec}^2 x - \cot x \operatorname{cosec} x$$
$$= -\operatorname{cosec} x(\operatorname{cosec} x + \cot x)$$
$$= -\operatorname{cosec} x \, g(x)$$

and this can be rewritten as

$$\operatorname{cosec} x = -\frac{g'(x)}{g(x)}.$$

We can now write

$$\int \operatorname{cosec} x \, dx = \int -\frac{g'(x)}{g(x)} \, dx$$
$$= -\ln|g(x)| + c$$
$$= -\ln|\operatorname{cosec} x + \cot x| + c.$$

The four results

$$\int \tan x \, dx = \ln|\sec x| + c$$

$$\int \cot x \, dx = \ln|\sin x| + c$$

$$\int \sec x \, dx = \ln|\sec x + \tan x| + c$$

$$\int \operatorname{cosec} x \, dx = -\ln|\operatorname{cosec} x + \cot x| + c$$

are given in the formula booklet – make sure you know where to find them!

Having studied this chapter you should know how to

- use the result $\int \frac{f'(x)}{f(x)} \, dx = \ln(|f(x)|) + c$

- find values P and Q so that $\dfrac{Ax + B}{(x + a)(x + b)} \equiv \dfrac{P}{x + a} + \dfrac{Q}{x + b}$

- use integration by substitution to simplify a variety of integrals

REVISION EXERCISE

1 A curve has equation $y = \dfrac{x}{x^2 + 4}$.

 i) Prove that the curve has a maximum point at $\left(2, \dfrac{1}{4}\right)$ and a minimum point at $\left(-2, -\dfrac{1}{4}\right)$ and no other stationary points.

 ii) Sketch the curve.

 iii) Find the area of the closed region bounded by the curve, the x-axis and the line $x = 2$.

2 Use integration by substitution to evaluate $\displaystyle\int_0^3 \dfrac{x}{\sqrt{16 + x^2}}\,dx$.

3 Express $\dfrac{11 - 12x}{(1 - 3x)(1 + 4x)}$ as a sum of partial fractions and hence prove that

$$\int \dfrac{11 - 12x}{(1 - 3x)(1 + 4x)}\,dx = \ln\left(\dfrac{(1 + 4x)^2}{|1 - 3x|}\right) + c.$$

4 Evaluate $\displaystyle\int_0^{\frac{1}{4}\pi} \dfrac{\sec^2 x}{1 + 4\tan x}\,dx$.

5 Use a substitution to find $\displaystyle\int e^{2x}\left(1 + 3e^{2x}\right)^2 dx$.

6 Evaluate $\displaystyle\int_0^3 \dfrac{x}{(x + 3)^2}\,dx$.

7 Prove that $\displaystyle\int_0^2 \dfrac{2x + 3}{(x + 1)(2x + 1)}\,dx = \ln\left(\dfrac{25}{3}\right)$.

8 Evaluate

 a) $\displaystyle\int_0^{\frac{1}{4}\pi} \dfrac{\cos 2x}{3 + 2\sin 2x}\,dx$ **b)** $\displaystyle\int_0^{\frac{1}{4}\pi} \dfrac{\cos 2x}{\sqrt{3 + 2\sin 2x}}\,dx$

9 The portion of the curve $y = \dfrac{\sqrt{\sin x}}{1 + \cos x}$ between $x = 0$ and $x = \dfrac{1}{2}\pi$ is rotated completely about the x-axis. Prove that the resulting solid has volume $\dfrac{3}{2}\pi$ units^3.

10 Evaluate

 a) $\displaystyle\int_0^3 \dfrac{x}{9 + x^2}\,dx$ **b)** $\displaystyle\int_0^3 \dfrac{x}{(9 + x^2)^2}\,dx$

3 The Binomial Expansion

The purpose of this chapter is to enable you to

- find and use the expansion of $(1 + z)^n$ where n is a real number

In module C2, we have seen that if n is a positive integer then

$$(1 + x)^n = 1 +\ _nC_1\, x +\ _nC_2\, x^2 +\ _nC_3\, x^3 + \cdots +\ _nC_r\, x^r + \cdots + x^n$$

or

$$(1 + x)^n = 1 + nx + \frac{n(n - 1)}{2!}\, x^2 + \frac{n(n - 1)(n - 2)}{3!}\, x^3 + \cdots + \frac{n(n - 1) \ldots (n - r + 1)}{r!}\, x^r + \cdots + x^n$$

and that

i) this expansion finishes with the x^n term;
ii) this expansion is valid for any value of x.

We now wish to investigate the expansion of $(1 + x)^n$ in cases where n is not a positive integer. We want to know whether the expansion

> Note that if n is not a positive whole number then this expansion will have an infinite number of terms.

$$(1 + x)^n = 1 + nx + \frac{n(n - 1)}{2!}\, x^2 + \frac{n(n - 1)(n - 2)}{3!}\, x^3 + \cdots + \frac{n(n - 1) \ldots (n - r + 1)}{r!}\, x^r + \cdots$$

is valid in cases when n is **not** a positive integer.

Investigating the Expansion for $(1 + x)^{\frac{1}{2}}$

With $n = \frac{1}{2}$ the expansion

$$(1 + x)^n = 1 + nx + \frac{n(n - 1)}{2!}\, x^2 + \frac{n(n - 1)(n - 2)}{3!}\, x^3 + \cdots + \frac{n(n - 1) \ldots (n - r + 1)}{r!}\, x^r + \cdots$$

becomes

$$(1 + x)^{\frac{1}{2}} = 1 + \frac{1}{2}x + \frac{\frac{1}{2}\left(\frac{1}{2} - 1\right)}{2!}\, x^2 + \frac{\frac{1}{2}\left(\frac{1}{2} - 1\right)\left(\frac{1}{2} - 2\right)}{3!}\, x^3 + \frac{\frac{1}{2}\left(\frac{1}{2} - 1\right)\left(\frac{1}{2} - 2\right)\left(\frac{1}{2} - 3\right)}{4!}\, x^4 + \cdots$$

which simplifies to

$$(1 + x)^{\frac{1}{2}} = 1 + \frac{1}{2}x - \frac{1}{8}x^2 + \frac{1}{16}x^3 - \frac{5}{128}x^4 + \cdots .$$

To investigate whether this expansion is valid, the graph of $y = (1 + x)^{\frac{1}{2}}$ will be compared with the graph obtained from the first few terms of the expansion.

The first diagram shows the graphs of $y = (1 + x)^{\frac{1}{2}}$ and $y = 1 + \dfrac{1}{2}x$.

We can see that for values of x between approximately -0.2 and 0.3 the values of $(1 + x)^{\frac{1}{2}}$ and $1 + \dfrac{1}{2}x$ appear to be close to each other.

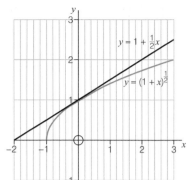

The second diagram shows the graphs of $y = (1 + x)^{\frac{1}{2}}$ and $y = 1 + \dfrac{1}{2}x - \dfrac{1}{8}x^2$.

This time, the values of $(1 + x)^{\frac{1}{2}}$ and $1 + \dfrac{1}{2}x - \dfrac{1}{8}x^2$ appear to be close to each other for values of x between approximately -0.4 and 0.5.

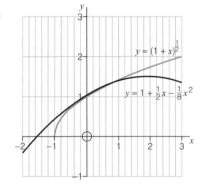

The third diagram shows the graphs of $y = (1 + x)^{\frac{1}{2}}$ and $y = 1 + \dfrac{1}{2}x - \dfrac{1}{8}x^2 + \dfrac{1}{16}x^3$.

The values of $(1 + x)^{\frac{1}{2}}$ and $1 + \dfrac{1}{2}x - \dfrac{1}{8}x^2 + \dfrac{1}{16}x^3$ appear to be close to each other for values of x between approximately -0.5 and 0.7.

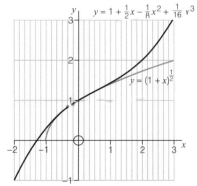

The final diagram shows the graph of $y = (1 + x)^{\frac{1}{2}}$ and the graph of the function consisting of the first six terms of the expansion:

$$y = 1 + \frac{1}{2}x - \frac{1}{8}x^2 + \frac{1}{16}x^3 - \frac{5}{128}x^4 + \frac{7}{256}x^5.$$

In this case there appears to be very close agreement for values of x between approximately -0.7 and 0.9.

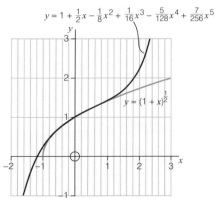

EXERCISE 1

Make sure you work through this exercise before proceeding to the next section.

A graphical calculator or computer graph-drawing package should be used to produce the graphs.

1 Consider the expansion of $(1+x)^{-2}$.

i) Write down and simplify the first six terms in the expansion

$$(1+x)^n = 1 + nx + \frac{n(n-1)}{2!}x^2 + \frac{n(n-1)(n-2)}{3!}x^3 + \cdots + \frac{n(n-1)\ldots(n-r+1)}{r!}x^r + \ldots$$

when $n = -2$.

ii) Using scales of $-2 \leqslant x \leqslant 2$ and $-10 \leqslant y \leqslant 10$, investigate how the graph of $y = (1+x)^{-2}$ compares with the graphs of

a) $y = 1 - 2x$

b) $y = 1 - 2x + 3x^2$

c) $y = 1 - 2x + 3x^2 - 4x^3$

d) $y = 1 - 2x + 3x^2 - 4x^3 + 5x^4$

iii) Continue the expansion up to the x^9 term.
How good is the expansion now?

2 Consider the expansion of $(1+x)^{-\frac{1}{4}}$.

i) Write down and simplify the first six terms in the expansion

$$(1+x)^n = 1 + nx + \frac{n(n-1)}{2!}x^2 + \frac{n(n-1)(n-2)}{3!}x^3 + \cdots + \frac{n(n-1)\ldots(n-r+1)}{r!}x^r + \ldots$$

when $n = -\frac{1}{4}$.

ii) Using scales of $-2 \leqslant x \leqslant 2$ and $-2 \leqslant y \leqslant 2$, investigate graphically the validity of this expansion.

3 Consider the expansion of $(1+x)^{\frac{3}{2}}$.

i) Write down and simplify the first six terms in the expansion

$$(1+x)^n = 1 + nx + \frac{n(n-1)}{2!}x^2 + \frac{n(n-1)(n-2)}{3!}x^3 + \cdots + \frac{n(n-1)\ldots(n-r+1)}{r!}x^r + \ldots$$

when $n = 1.5$.

ii) Using scales of $-2 \leqslant x \leqslant 2$ and $-1 \leqslant y \leqslant 6$, investigate graphically the validity of this expansion.

4 **Review:** Look back at the example in the previous section and the three cases considered in this exercise.
For what values of x does it look as if the expansion

$$(1+x)^n = 1 + nx + \frac{n(n-1)}{2!}x^2 + \frac{n(n-1)(n-2)}{3!}x^3 + \cdots + \frac{n(n-1)\ldots(n-r+1)}{r!}x^r + \ldots$$

is valid?

The Expansion of $(1 + x)^n$

The results of the previous section and the questions of Exercise 1 should suggest that

If n is not a positive integer then the series

$$1 + nx + \frac{n(n-1)}{2!}x^2 + \frac{n(n-1)(n-2)}{3!}x^3 + \cdots + \frac{n(n-1)\ldots(n-r+1)}{r!}x^r + \ldots$$

converges to the value of $(1 + x)^n$
provided $-1 < x < 1$.

The series $1 + nx + \frac{n(n-1)}{2!}x^2 + \frac{n(n-1)(n-2)}{3!}x^3 + \cdots + \frac{n(n-1)\ldots(n-r+1)}{r!}x^r + \cdots$ is called

the binomial expansion of $(1 + x)^n$.

If n is **not** a positive integer then the series converges to the value of $(1 + x)^n$ only if the value of x lies between -1 and 1.

Recalling that

$r!$ is shorthand for $1.2.3 \ldots r$

and that

the inequality $-1 < x < 1$ can be written as $|x| < 1$

the fundamental result can be rewritten as

For any real number n

$$(1+x)^n = 1 + nx + \frac{n(n-1)}{1.2}x^2 + \frac{n(n-1)(n-2)}{1.2.3}x^3 + \cdots + \frac{n(n-1)\ldots(n-r+1)}{1.2.3\ldots r}x^r + \ldots$$

provided $|x| < 1$.

EXAMPLE 1

a) Find the first three terms in the binomial expansion for $(1 + 9x)^{\frac{1}{3}}$, stating the values of x for which the series is valid.

b) By putting $x = 0.001$ find an estimate for the value of $\sqrt[3]{1009}$.

SOLUTION

a) Starting with the binomial expansion of $(1 + z)^n$

$$(1+z)^n = 1 + nz + \frac{n(n-1)}{1.2}z^2 + \ldots \qquad \text{for} \quad -1 < z < 1$$

and putting $n = \frac{1}{3}$

$$(1+z)^{\frac{1}{3}} = 1 + \frac{1}{3}z + \frac{\frac{1}{3}\left(\frac{1}{3}-1\right)}{2!}z^2 + \ldots$$

$$= 1 + \frac{1}{3}z - \frac{1}{9}z^2 + \ldots \qquad \text{for} \quad -1 < z < 1.$$

EXAMPLE 1 (continued)

Putting $z = 9x$ gives

$$(1 + 9x)^{\frac{1}{3}} = 1 + \frac{1}{3}(9x) - \frac{1}{9}(9x)^2 + \dots \qquad \text{for} \quad -1 < 9x < 1$$

$$\Rightarrow \quad (1 + 9x)^{\frac{1}{3}} = 1 + 3x - 9x^2 + \dots \qquad \text{for} \quad -\frac{1}{9} < x < \frac{1}{9}.$$

b) Putting $x = 0.001$ gives

$$(1 + 0.009)^{\frac{1}{3}} \approx 1 + 3 \times 0.001 - 9 \times 0.000001 = 1.002991$$

$$\Rightarrow \quad \sqrt[3]{1.009} \approx 1.002991.$$

Now

$$\sqrt[3]{1009} = \sqrt[3]{1000 \times 1.009} = \sqrt[3]{1000} \times \sqrt[3]{1.009} = 10 \times \sqrt[3]{1.009}$$

$$\approx 10 \times 1.002991$$

$$= 10.02991.$$

EXAMPLE 2

Find the first four terms in the expansion of $\sqrt{\dfrac{1 + 4x}{1 - 2x}}$ and state the values of x for which the expansion is valid.

First notice that

$$\sqrt{\frac{1 + 4x}{1 - 2x}} = (1 + 4x)^{1/2}(1 - 2x)^{-1/2}.$$

By the binomial expansion,

$$(1 + z)^{1/2} = 1 + \frac{1}{2}z + \frac{\frac{1}{2}\left(\frac{1}{2} - 1\right)}{2!}z^2 + \frac{\frac{1}{2}\left(\frac{1}{2} - 1\right)\left(\frac{1}{2} - 2\right)}{3!}z^3 + \dots$$

$$= 1 + \frac{1}{2}z - \frac{1}{8}z^2 + \frac{1}{16}z^3 + \dots \qquad \text{provided} \quad -1 < z < 1.$$

Now, putting $z = 4x$, we obtain

$$(1 + 4x)^{1/2} = 1 + \frac{1}{2}(4x) - \frac{1}{8}(4x)^2 + \frac{1}{16}(4x)^3 + \dots$$

$$= 1 + 2x - 2x^2 + 4x^3 + \dots$$

provided $-1 < 4x < 1$ or, equivalently, $-\frac{1}{4} < x < \frac{1}{4}$.

Similarly, by the binomial expansion,

$$(1 + z)^{-1/2} = 1 + \frac{-1}{2}z + \frac{\frac{-1}{2}\left(\frac{-1}{2} - 1\right)}{2!}z^2 + \frac{\frac{-1}{2}\left(\frac{-1}{2} - 1\right)\left(\frac{-1}{2} - 2\right)}{3!}z^3 + \dots$$

$$= 1 - \frac{1}{2}z + \frac{3}{8}z^2 - \frac{5}{16}z^3 + \dots \qquad \text{provided} \quad -1 < z < 1.$$

EXAMPLE 2 (continued)

Now, putting $z = -2x$, we obtain

$$(1 - 2x)^{-1/2} = 1 - \frac{1}{2}(-2x) + \frac{3}{8}(-2x)^2 - \frac{5}{16}(-2x)^3 + \ldots$$

$$= 1 + x + \frac{3}{2}x^2 + \frac{5}{2}x^3 + \ldots$$

provided $-1 < -2x < 1$ or, equivalently, $-\frac{1}{2} < x < \frac{1}{2}$.

> Dividing the inequality
> $$-1 < -2x < 1$$
> by -2 gives
> $$\tfrac{1}{2} > x > -\tfrac{1}{2}$$
> since dividing by a negative number reverses all the inequalities. This inequality can be written
> $$-\tfrac{1}{2} < x < \tfrac{1}{2}.$$

Now

$$\sqrt{\frac{1 + 4x}{1 - 2x}} = (1 + 4x)^{1/2}(1 - 2x)^{-1/2}$$

$$= (1 + 2x - 2x^2 + 4x^3 + \ldots)\left(1 + x + \frac{3}{2}x^2 + \frac{5}{2}x^3 + \ldots\right)$$

$$= 1 + x + \frac{3}{2}x^2 + \frac{5}{2}x^3 + \cdots + 2x + 2x^2 + 3x^3 + \cdots - 2x^2 - 2x^3 + \cdots + 4x^3 + \ldots$$

$$= 1 + 3x + \frac{3}{2}x^2 + \frac{15}{2}x^3 + \ldots.$$

> Multiplying out the brackets and ignoring terms in x^4 and higher powers of x.

The series will only be valid if $-\frac{1}{4} < x < \frac{1}{4}$.

> The series for $(1 + 4x)^{1/2}$ is only valid for $-\frac{1}{4} < x < \frac{1}{4}$ and the series for $(1 - 2x)^{-1/2}$ is only valid for $-\frac{1}{2} < x < \frac{1}{2}$ so the series for $(1 + 4x)^{1/2}(1 - 2x)^{-1/2}$ is only valid for values of x which satisfy **both** inequalities: i.e. values of x between $-\frac{1}{4}$ and $\frac{1}{4}$.

The Expansion of $(a + x)^n$

Expansions for expressions of the form $(a + x)^n$ can be obtained by first writing

$$(a + x)^n = \left(a\left(1 + \frac{x}{a}\right)\right)^n = a^n\left(1 + \frac{x}{a}\right)^n.$$

EXAMPLE 3

Find the first four non-zero terms in the expansion of $\dfrac{1}{4 + x^2}$ and state the values of x for which the series is valid.

First observe that

$$\frac{1}{4 + x^2} = (4 + x^2)^{-1}.$$

Now

$$\frac{1}{4 + x^2} = (4 + x^2)^{-1} = \left(4\left(1 + \frac{x^2}{4}\right)\right)^{-1} = 4^{-1}\left(1 + \frac{x^2}{4}\right)^{-1} = \frac{1}{4}\left(1 + \frac{x^2}{4}\right)^{-1}.$$

EXAMPLE 3 (continued)

By the binomial expansion,

$$(1+z)^{-1} = 1 + (-1)z + \frac{-1(-1-1)}{2!}z^2 + \frac{-1(-1-1)(-1-2)}{3!}z^3 + \dots$$

$$= 1 - z + z^2 - z^3 + \dots \qquad \text{provided} \quad -1 < z < 1.$$

Now, putting $z = \dfrac{x^2}{4}$ we obtain

$$\left(1 + \frac{x^2}{4}\right)^{-1} = 1 - \left(\frac{x^2}{4}\right) + \left(\frac{x^2}{4}\right)^2 - \left(\frac{x^2}{4}\right)^3 + \dots$$

$$= 1 - \frac{1}{4}x^2 + \frac{1}{16}x^4 - \frac{1}{64}x^6 + \dots$$

provided $-1 < \dfrac{x^2}{4} < 1$ or, equivalently, $-2 < x < 2$.

So

$$\frac{1}{4+x^2} = \frac{1}{4}\left(1 + \frac{x^2}{4}\right)^{-1}$$

$$= \frac{1}{4}\left(1 - \frac{1}{4}x^2 + \frac{1}{16}x^4 - \frac{1}{64}x^6 + \dots\right)$$

$$= \frac{1}{4} - \frac{1}{16}x^2 + \frac{1}{64}x^4 - \frac{1}{256}x^6 + \dots$$

provided $-2 < x < 2$.

> $-1 < \dfrac{x^2}{4} < 1 \Rightarrow -4 < x^2 < 4.$
>
> The inequality $-4 < x^2$ is satisfied by **all** real numbers since a square number is always positive.
> We therefore just need to ensure that $x^2 < 4$.
>
> The solution of the inequality $x^2 < 4$ is $-2 < x < 2$.

EXAMPLE 4

i) Obtain the expansions, up to the x^3 term, of

    a) $\dfrac{1}{1+4x}$ **b)** $\dfrac{1}{2-x}$

ii) Write $\dfrac{2+35x}{(1+4x)(2-x)}$ as a sum of partial fractions.

iii) Hence find the expansion, up to the x^3 term, of $\dfrac{2+35x}{(1+4x)(2-x)}$.

i) a) Using the binomial expansion

$$(1+z)^{-1} = 1 + (-1)z + \frac{-1(-1-1)}{2!}z^2 + \frac{-1(-1-1)(-1-2)}{3!}z^3 + \dots$$

$$= 1 - z + z^2 - z^3 + \dots.$$

Putting $z = 4x$ gives

$$\frac{1}{1+4x} = (1+4x)^{-1}$$

$$= 1 - (4x) + (4x)^2 - (4x)^3 + \dots$$

$$= 1 - 4x + 16x^2 - 64x^3.$$

EXAMPLE 4 (continued)

b) $\dfrac{1}{2-x} = (2-x)^{-1} = \left(2\left(1-\dfrac{x}{2}\right)\right)^{-1} = 2^{-1}\left(1-\dfrac{x}{2}\right)^{-1} = \dfrac{1}{2}\left(1-\dfrac{x}{2}\right)^{-1}.$

Putting $z = -\dfrac{1}{2}x$ into the expansion of $(1+z)^{-1}$ gives

$$\dfrac{1}{2-x} = \dfrac{1}{2}\left(1-\dfrac{x}{2}\right)^{-1}$$

$$= \dfrac{1}{2}\left[1 - \left(-\dfrac{1}{2}x\right) + \left(-\dfrac{1}{2}x\right)^2 - \left(-\dfrac{1}{2}x\right)^3 + \ldots\right]$$

$$= \dfrac{1}{2}\left[1 + \dfrac{1}{2}x + \dfrac{1}{4}x^2 + \dfrac{1}{8}x^3 + \ldots\right]$$

$$= \dfrac{1}{2} + \dfrac{1}{4}x + \dfrac{1}{8}x^2 + \dfrac{1}{16}x^3 + \ldots.$$

ii) If

$$\dfrac{2+35x}{(1+4x)(2-x)} \equiv \dfrac{P}{1+4x} + \dfrac{Q}{2-x}$$

then multiplying through by $(1+4x)(2-x)$ gives

$$2 + 35x \equiv P(2-x) + Q(1+4x).$$

Putting $x = 2$ gives

$$72 = 0P + 9Q \quad \Rightarrow \quad Q = 8.$$

Putting $x = -\dfrac{1}{4}$ gives

$$-\dfrac{27}{4} = \dfrac{9}{4}P \quad \Rightarrow \quad P = -3.$$

This means that

$$\dfrac{2+35x}{(1+4x)(2-x)} = \dfrac{-3}{1+4x} + \dfrac{8}{2-x}.$$

iii) Using parts (i) and (ii) we can write

$$\dfrac{2+35x}{(1+4x)(2-x)} = \dfrac{-3}{1+4x} + \dfrac{8}{2-x}$$

$$= 8 \times \dfrac{1}{2-x} - 3 \times \dfrac{1}{1+4x}$$

$$= 8 \times \left(\dfrac{1}{2} + \dfrac{1}{4}x + \dfrac{1}{8}x^2 + \dfrac{1}{16}x^3 + \ldots\right) - 3 \times \left(1 - 4x + 16x^2 - 64x^3 + \ldots\right)$$

$$= 4 + 2x + x^2 + \dfrac{1}{2}x^3 - 3 + 12x - 48x^2 + 192x^3 + \ldots$$

$$= 1 + 14x - 47x^2 + \dfrac{385}{2}x^3 + \ldots.$$

EXERCISE 2

For questions 1–6, write down the first four terms in the binomial expansions for each of the following expressions and state the values of x for which they are valid:

1 $(1 + x)^{-1/2}$

2 $(1 + x)^{-3}$

3 $\dfrac{1}{1 - 4x}$

4 $\dfrac{1}{(2 + x)^2}$

5 $\sqrt{9 + x}$

6 $(2 - 3x)^{-2}$

7 Write out the binomial expansion for $(1 - 2x)^{-2}$ up to the x^4 term.

Hence find the expansion for $\dfrac{2 + 3x}{(1 - 2x)^2}$ up to the x^4 term and state the values of x for which the series is valid.

8 Find an expression in ascending powers of x, up to the x^3 term, for $\sqrt{\dfrac{4 + x^2}{1 - 3x}}$.

9 Find the expansion of $\sqrt{25 + x}$ up to the x^2 term and state the values of x for which the expansion is valid.

Hence prove that $\sqrt{27} \approx 5\dfrac{49}{250}$.

10 a) Write $\dfrac{4 - x}{(1 - x)(1 + 2x)}$ as a sum of partial fractions.

b) Find the binomial expansions, up to the x^3 term, for **i)** $\dfrac{1}{1 - x}$ **ii)** $\dfrac{1}{1 + 2x}$

c) Hence find the expansion, up to the x^3 term, for $\dfrac{4 - x}{(1 - x)(1 + 2x)}$.

11 Show that the first three terms in the expansion, in ascending powers of x of $(1 + 8x)^{1/4}$ are the same as the first three terms in the expansion of $\dfrac{1 + 5x}{1 + 3x}$.

Use the corresponding approximation, $(1 + 8x)^{1/4} \approx \dfrac{1 + 5x}{1 + 3x}$, to find integers p and q such that $\sqrt[4]{1.16} \approx \dfrac{p}{q}$.

Having studied this chapter you should know how to

● use the binomial expansion

$$(1 + x)^n = 1 + nx + \frac{n(n - 1)}{2!}x^2 + \frac{n(n - 1)(n - 2)}{3!}x^3 + \cdots + \frac{n(n - 1) \ldots (n - r + 1)}{r!}x^r + \cdots$$

and know that if n is not a positive integer then the expansion is only valid for $-1 < x < 1$

● obtain the expansion of $(a + x)^n$ by first writing

$$(a + x)^n = \left(a\left(1 + \frac{x}{a}\right) \right)^n = a^n \left(1 + \frac{x}{a}\right)^n$$

and then expanding $\left(1 + \frac{x}{a}\right)^n$

REVISION EXERCISE

1 Find the first four terms of the expansion in ascending powers of x of $\dfrac{1}{(1 + 4x)^2}$.
State the values of x for which this expansion is valid.

2 Find the first three terms of the expansion in ascending powers of x of $\sqrt{25 + x}$ and state the values of x for which this expansion is valid.
By letting $x = 0.1$, and making your method clear, find an approximation for $\sqrt{2510}$.

3 Write $\dfrac{12x + 8}{(1 + 3x)(1 - x)}$ as a sum of two partial fractions and hence find the first four terms of the expansion in ascending powers of x of $\dfrac{12x + 8}{(1 + 3x)(1 - x)}$. State the values of x for which this expansion is valid.

4 Find the first four terms in the expansion in ascending powers of x of $(1 + 8x)^{\frac{1}{4}}$.
In the expansion of $(1 + px)(1 + 8x)^{\frac{1}{4}}$ the coefficient of x^2 is -16. Find the possible values of the constant p.

5 In the expansion of $(1 + qx)^{-4}$ in ascending powers of x, the coefficient of x^3 is four times the coefficient of x^2. Determine the value of q and hence write down the first four terms of the expansion of $(1 + qx)^{-4}$. For what values of x is the expansion valid?

6 a) Find the expansions, in ascending powers of x, up to the x^4 term of

   i) $\sqrt{1 + px}$ **ii)** $\dfrac{1}{\sqrt{1 + qx^2}}$

b) The first three terms in the expansion in ascending powers of x of $\sqrt{\dfrac{1 + px}{1 + qx^2}}$ are $1 + x + \dfrac{3}{2}x^2$.

   i) Find the values of the constants p and q.

   ii) Determine the next two terms in the expansion of $\sqrt{\dfrac{1 + px}{1 + qx^2}}$.

7 i) Determine the first three terms in the expansion in ascending powers of x of $(27 + x)^{\frac{1}{3}}$. State the set of values of x for which this expansion is valid.

   ii) Deduce that $\sqrt[3]{28} \approx 3\dfrac{80}{2187}$.

8 a) In the expansion of $\sqrt{9 + ax}$ in ascending powers of x, the coefficient of x^2 is $-\dfrac{2}{3}$.
Determine the possible values of the constant a.

b) The first three terms in the expansion of $(1 + px)^\alpha$ are $1 - 10x + 75x^2$. Determine the values of the constants p and α.

4 Vector Geometry

The purpose of this chapter is to enable you to

- use vector notation

- perform vector arithmetic and interpret the results geometrically

- calculate the magnitude of a vector

- use the vector equation of a straight line for lines in two- and three-dimensional space

- determine whether two lines are parallel, intersecting or skew

- calculate the scalar product of two vectors and use it to find angles between vectors and angles between lines

Basic Concepts

A vector is something with both size (or magnitude) and direction.

In pure maths, we have already used vectors in two-dimensional co-ordinate geometry and to describe translations. Force, velocity and acceleration are all examples of vectors that you may have met in Mechanics modules.

In co-ordinate geometry, a vector is often thought of as the directed line segment joining two points together.

For example, if A(2, 2) and B(7, 5) then the vector $\overrightarrow{AB}$ is the line segment joining A to B and is often represented by $\begin{pmatrix} 5 \\ 3 \end{pmatrix}$.

Note that 5 is called the x-component of the vector $\begin{pmatrix} 5 \\ 3 \end{pmatrix}$ and 3 is called its y-component.

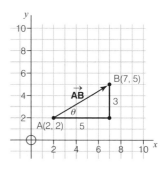

The length or **magnitude** of the vector $\overrightarrow{AB}$ is the length of the line segment AB and is easily calculated using Pythagoras's theorem:

$$AB = \sqrt{5^2 + 3^2} = \sqrt{34}.$$

The **direction** of a vector can be found using right-angled triangle trigonometry. For example, the vector $\overrightarrow{AB}$ makes an angle θ with the positive x-direction where

$$\tan \theta = \frac{3}{5}$$

$$\Rightarrow \quad \theta = \tan^{-1}\left(\frac{3}{5}\right) \approx 31°.$$

Sometimes we want to consider vectors without reference to the points that they link up. In this case we write the vector as $\underline{a} = \begin{pmatrix} p \\ q \end{pmatrix}$ and denote its length by $a = |\underline{a}| = \sqrt{p^2 + q^2}$.

> The vector $\begin{pmatrix} p \\ q \end{pmatrix}$ has length or magnitude $\sqrt{p^2 + q^2}$.

Addition of Vectors

If A(2, 2), B(7, 5) and C(8, 9) then $\overrightarrow{AB} = \begin{pmatrix} 5 \\ 3 \end{pmatrix}$, $\overrightarrow{BC} = \begin{pmatrix} 1 \\ 4 \end{pmatrix}$ and $\overrightarrow{AC} = \begin{pmatrix} 6 \\ 7 \end{pmatrix}$.

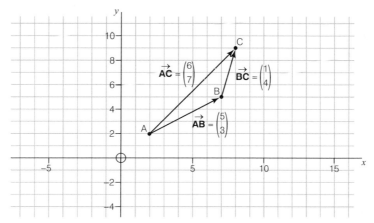

Moving from A to B and then moving from B to C gives the same overall effect as moving directly from A to C. This is expressed as $\overrightarrow{AB} + \overrightarrow{BC} = \overrightarrow{AC}$.

The vectors $\overrightarrow{AB}$ and $\overrightarrow{BC}$ are added arithmetically by adding the x-components and then adding the y-components to obtain

$$\begin{pmatrix} 5 \\ 3 \end{pmatrix} + \begin{pmatrix} 1 \\ 4 \end{pmatrix} = \begin{pmatrix} 6 \\ 7 \end{pmatrix}$$

but the vector $\begin{pmatrix} 6 \\ 7 \end{pmatrix}$ is $\overrightarrow{AC}$.

The vector $\underline{p} + \underline{q}$ can thus be found arithmetically by adding the x-components and then adding the y-components.

Geometrically, $\underline{p} + \underline{q}$ is equivalent to first going along $\underline{p}$ and then going along $\underline{q}$.

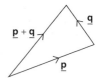

Multiplication of a Vector by a Scalar

If $k > 0$ then the vector $k\underline{a}$ is a vector in the same direction as $\underline{a}$ and k times as long.
If $k < 0$ then the vector $k\underline{a}$ is a vector in the opposite direction to $\underline{a}$ and $|k|$ times as long.

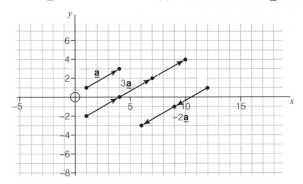

If $\underline{a} = \begin{pmatrix} 3 \\ 2 \end{pmatrix}$ then $3\underline{a} = \begin{pmatrix} 9 \\ 6 \end{pmatrix} = \begin{pmatrix} 3 \times 3 \\ 3 \times 2 \end{pmatrix}$ and $-2\underline{a} = \begin{pmatrix} -6 \\ -4 \end{pmatrix} = \begin{pmatrix} -2 \times 3 \\ -2 \times 2 \end{pmatrix}$.

Arithmetically, to multiply a vector by a scalar, just multiply each of the components by the scalar.

Subtraction of Vectors

Since

$$\underline{a} - \underline{b} = \underline{a} + (-\underline{b})$$

the vector $\underline{a} - \underline{b}$ can be regarded geometrically as the effect of going along $\underline{a}$ and then going along $(-\underline{b})$.

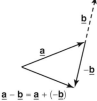

$\underline{a} - \underline{b} = \underline{a} + (-\underline{b})$

If $\underline{a} = \begin{pmatrix} 5 \\ 2 \end{pmatrix}$ and $\underline{b} = \begin{pmatrix} 2 \\ 4 \end{pmatrix}$ then arithmetically we can write

$$\underline{a} - \underline{b} = \underline{a} + (-\underline{b}) = \begin{pmatrix} 5 \\ 2 \end{pmatrix} + \left[-\begin{pmatrix} 2 \\ 4 \end{pmatrix} \right] = \begin{pmatrix} 5 \\ 2 \end{pmatrix} + \begin{pmatrix} -2 \\ -4 \end{pmatrix} = \begin{pmatrix} 5 + (-2) \\ 2 + (-4) \end{pmatrix} = \begin{pmatrix} 5 - 2 \\ 2 - 4 \end{pmatrix} = \begin{pmatrix} 3 \\ 2 \end{pmatrix}$$

and it is now evident that vectors can be subtracted arithmetically by simply performing subtractions on each component.

EXAMPLE 1

If $\underline{a} = \begin{pmatrix} 4 \\ -1 \end{pmatrix}$ and $\underline{b} = \begin{pmatrix} -2 \\ 5 \end{pmatrix}$, evaluate

i) $3\underline{a} + 2\underline{b}$ ii) $\dfrac{1}{2}\underline{a} - \dfrac{3}{2}\underline{b}$

SOLUTION

i) $3\underline{a} + 2\underline{b} = 3\begin{pmatrix} 4 \\ -1 \end{pmatrix} + 2\begin{pmatrix} -2 \\ 5 \end{pmatrix} = \begin{pmatrix} 12 \\ -3 \end{pmatrix} + \begin{pmatrix} -4 \\ 10 \end{pmatrix} = \begin{pmatrix} 8 \\ 7 \end{pmatrix}$.

ii) $\dfrac{1}{2}\underline{a} - \dfrac{3}{2}\underline{b} = \dfrac{1}{2}\begin{pmatrix} 4 \\ -1 \end{pmatrix} - \dfrac{3}{2}\begin{pmatrix} -2 \\ 5 \end{pmatrix} = \begin{pmatrix} 2 \\ -0.5 \end{pmatrix} - \begin{pmatrix} -3 \\ 7.5 \end{pmatrix} = \begin{pmatrix} 5 \\ -8 \end{pmatrix}$.

EXAMPLE 2

The diagram shows two vectors. Draw the vectors $\mathbf{p} + 2\mathbf{q}$ and $3\mathbf{p} - 2\mathbf{q}$.

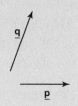

SOLUTION

$\mathbf{p} + 2\mathbf{q}$ is the same as going along $\mathbf{p}$ and then going along $2\mathbf{q}$.
$3\mathbf{p} - 2\mathbf{q}$ is the same as going along $3\mathbf{p}$ and then going along $-2\mathbf{q}$.

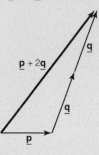

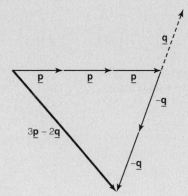

EXAMPLE 3

$\mathbf{a}$ and $\mathbf{b}$ are two vectors of length 5 and 3 respectively. The direction of $\mathbf{b}$ makes an angle of $60°$ with that of $\mathbf{a}$. Find the length and direction of the vector $2\mathbf{a} + \mathbf{b}$.

SOLUTION

Start by drawing a diagram of the vectors.

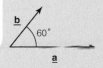

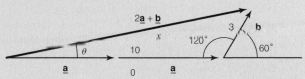

Using the cosine rule:

$$x^2 = 10^2 + 3^2 - 2 \times 10 \times 3 \times \cos 120°$$
$$\Rightarrow \quad x^2 = 139$$
$$\Rightarrow \quad x = \sqrt{139}.$$

If θ is the angle between $2\mathbf{a} + \mathbf{b}$ and $\mathbf{a}$ then the sine rule gives

$$\frac{\sin \theta}{3} = \frac{\sin 120°}{\sqrt{139}} \Rightarrow \theta = 12.7° \quad \text{(1 d.p.)}$$

The vector $2\mathbf{a} + \mathbf{b}$ has length $\sqrt{139}$ and acts at an angle of $12.7°$ to vector $\mathbf{a}$.

EXERCISE 1

1 Copy the diagram and on your diagram also show the vectors

i) $\underline{a} + \underline{b}$
ii) $\underline{a} - \underline{b}$
iii) $3\underline{a} + \underline{b}$
iv) $3\underline{a} - 4\underline{b}$

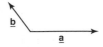

2 Copy the diagram and on your diagram also show the vectors

i) $3\underline{a}$
ii) $\underline{a} + \dfrac{1}{2}\underline{b}$
iii) $\dfrac{1}{2}\underline{a} - \underline{b}$

3 Evaluate

a) $4\begin{pmatrix} 2 \\ -1 \end{pmatrix} - 3\begin{pmatrix} -1 \\ 2 \end{pmatrix}$

b) $2\begin{pmatrix} 5 \\ 3 \end{pmatrix} + 5\begin{pmatrix} -1 \\ 1 \end{pmatrix}$

c) $\dfrac{1}{2}\begin{pmatrix} 3 \\ -6 \end{pmatrix} + \dfrac{3}{2}\begin{pmatrix} 1 \\ 4 \end{pmatrix}$

4 Copy the diagram of the triangle OAB.
The points P, Q and R are such that

$$\overrightarrow{OP} = \frac{1}{4}\underline{a} + \frac{3}{4}\underline{b}$$

$$\overrightarrow{OQ} = \frac{1}{2}\underline{a} + \frac{1}{2}\underline{b}$$

$$\overrightarrow{OR} = \frac{3}{4}\underline{a} + \frac{1}{4}\underline{b}$$

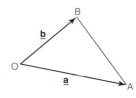

Mark the points P, Q and R on your diagram.

5 $\underline{a}$ and $\underline{b}$ are two vectors of length 7 and 4 respectively. The direction of $\underline{b}$ makes an angle of 75° with that of $\underline{a}$. Find the length of the vector $3\underline{a} - 2\underline{b}$ and find the angle that this vector makes with $\underline{a}$.

6 In the triangle OAB $\overrightarrow{OA} = \underline{a}$ and $\overrightarrow{OB} = \underline{b}$ and P is the point such that $\overrightarrow{AP} = \dfrac{4}{5}\overrightarrow{AB}$.
Find expressions, in terms of $\underline{a}$ and $\underline{b}$, for

a) $\overrightarrow{AB}$
b) $\overrightarrow{OP}$

If, moreover, $\underline{a}$ and $\underline{b}$ have lengths of 8 units and 5 units respectively and $\angle BOA = 45°$, calculate $\angle BOP$.

7 OABC is a trapezium in which BC is parallel to AO and three times as long. If $\overrightarrow{OA} = \underline{a}$ and $\overrightarrow{OC} = \underline{c}$, find $\overrightarrow{OB}$ in terms of $\underline{a}$ and $\underline{c}$.
E and F are the midpoints of AC and OB respectively. Find $\overrightarrow{OE}$ and $\overrightarrow{OF}$ in terms of $\underline{a}$ and $\underline{c}$.
Determine the nature of the quadrilateral OEFA.

The Position Vector of a Point

The **position vector** of a point is the vector describing the movement from the origin to the point.

If P is the point (6, 4) then the position vector of P is the vector $\overrightarrow{OP} = \begin{pmatrix} 6 \\ 4 \end{pmatrix}$.

The Vector Equation of a Line

The diagram shows the line $y = \dfrac{1}{2}x + 5$. Observe that the point A(0, 4) is on the line and that the vector $\begin{pmatrix} 4 \\ 2 \end{pmatrix}$ represents a movement along the line.

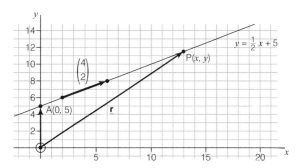

Let P(x, y) be a general point on the line and let **r** denote the position vector of P. We can write

$$\underline{r} = \overrightarrow{OP} = \overrightarrow{OA} + \overrightarrow{AP}$$

to describe the fact that to travel from O to P it is possible to move from O onto the line at A and then move along the line from A to P.

The movement from O to A is given by the vector $\begin{pmatrix} 0 \\ 5 \end{pmatrix}$.

The movement from A to P is a vector parallel to the vector $\begin{pmatrix} 4 \\ 2 \end{pmatrix}$ and must therefore be a multiple of $\begin{pmatrix} 4 \\ 2 \end{pmatrix}$. We can therefore write $\overrightarrow{AP} = t\begin{pmatrix} 4 \\ 2 \end{pmatrix}$, where t is a number, and deduce that

$$\underline{r} = \overrightarrow{OP} = \overrightarrow{OA} + \overrightarrow{AP} = \begin{pmatrix} 0 \\ 5 \end{pmatrix} + t\begin{pmatrix} 4 \\ 2 \end{pmatrix}.$$

The equation $\underline{r} = \begin{pmatrix} 0 \\ 5 \end{pmatrix} + t\begin{pmatrix} 4 \\ 2 \end{pmatrix}$ is a **vector equation** for the line $y = \dfrac{1}{2}x + 5$.

In looking at the equation $\underline{r} = \begin{pmatrix} 0 \\ 5 \end{pmatrix} + t\begin{pmatrix} 4 \\ 2 \end{pmatrix}$,

The choice of the letter t to denote the multiple of $\begin{pmatrix} 4 \\ 2 \end{pmatrix}$ is purely arbitrary. We could just have well used s or λ and obtained

$$\underline{r} = \begin{pmatrix} 0 \\ 5 \end{pmatrix} + s\begin{pmatrix} 4 \\ 2 \end{pmatrix} \text{ or } \underline{r} = \begin{pmatrix} 0 \\ 5 \end{pmatrix} + \lambda\begin{pmatrix} 4 \\ 2 \end{pmatrix}$$

as the vector equation of the line.

note that the vector $\begin{pmatrix} 0 \\ 5 \end{pmatrix}$ takes us **from the origin**

onto the line, whilst the vector $t\begin{pmatrix} 4 \\ 2 \end{pmatrix}$ **gives a movement along the line**. We say that the vector $\begin{pmatrix} 4 \\ 2 \end{pmatrix}$ is a **direction vector for the line**.

Notice that the vector equation of a line is not unique. For example, each of the following vector equations also represents the line $y = \frac{1}{2}x + 5$:

$$\mathbf{\underline{r}} = \begin{pmatrix} 0 \\ 5 \end{pmatrix} + t\begin{pmatrix} 4 \\ 2 \end{pmatrix}, \quad \mathbf{\underline{r}} = \begin{pmatrix} -2 \\ 6 \end{pmatrix} + t\begin{pmatrix} 4 \\ 2 \end{pmatrix}, \quad \mathbf{\underline{r}} = \begin{pmatrix} 0 \\ 5 \end{pmatrix} + s\begin{pmatrix} 2 \\ 1 \end{pmatrix}, \quad \mathbf{\underline{r}} = \begin{pmatrix} 8 \\ 9 \end{pmatrix} + \lambda\begin{pmatrix} -8 \\ -4 \end{pmatrix}.$$

The argument above can be generalised.

A line passes through a point A whose position vector is **a**. The line has direction vector **b**.

If P is an arbitrary point, with position vector **r**, on the line then

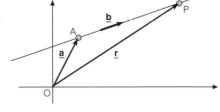

$$\mathbf{\underline{r}} = \overrightarrow{OP} = \overrightarrow{OA} + \overrightarrow{AP}.$$

The movement $\overrightarrow{OA}$ is given by the position vector, **a**, of A and the movement from A to P is a vector parallel to the vector **b** so we can therefore write $\overrightarrow{AP} = t\mathbf{\underline{b}}$. The equation of the line can therefore be written as

$$\mathbf{\underline{r}} = \mathbf{\underline{a}} + t\mathbf{\underline{b}}.$$

> The equation $\mathbf{\underline{r}} = \mathbf{\underline{a}} + t\mathbf{\underline{b}}$ represents a straight line which
>
> - passes through the point A whose position vector is **a**
> - has direction vector **b**

EXAMPLE 4

Find a vector equation for the line passing through the points C and D whose co-ordinates are (5, 7) and (7, 2).

$\overrightarrow{CD} = \begin{pmatrix} 2 \\ -5 \end{pmatrix}$ is a direction vector for the line.

The line passes through C(5, 7) whose position vector is $\begin{pmatrix} 5 \\ 7 \end{pmatrix}$.

A possible equation for the line is therefore $\mathbf{\underline{r}} = \begin{pmatrix} 5 \\ 7 \end{pmatrix} + t\begin{pmatrix} 2 \\ -5 \end{pmatrix}$.

Obtaining the Cartesian Equation of a Line from its Vector Equation

It is easy to move from the vector equation of a line back to the Cartesian equation of the line.

EXAMPLE 5

Find the Cartesian equation of the line $\mathbf{\underline{r}} = \begin{pmatrix} 5 \\ -1 \end{pmatrix} + t\begin{pmatrix} 1 \\ 3 \end{pmatrix}$.

EXAMPLE 5 (continued)

By putting $t = 0$ and $t = 1$ it can be seen that the line passes through the points $(5, -1)$ and $(6, 2)$.

$$\text{Gradient of line} = \frac{y\text{-step}}{x\text{-step}} = \frac{2 - (-1)}{6 - 5} = 3.$$

The line has gradient 3 and passes through $(5, -1)$ so its equation must be

$$y - (-1) = 3(x - 5)$$
$$\Rightarrow \quad y + 1 = 3x - 15$$
$$\Rightarrow \quad y = 3x - 16.$$

Recall that

$$y - y_1 = m(x - x_1)$$

is the equation of the straight line of gradient m passing through the point (x_1, y_1).

Finding the Point of Intersection of Two Lines from their Vector Equations

The **point of intersection** of two lines whose vector equations are known can easily be found without first finding the Cartesian equations of the lines.

EXAMPLE 6

Find the point of intersection of the lines

$$\mathbf{r_1} = \begin{pmatrix} 3 \\ -2 \end{pmatrix} + s\begin{pmatrix} 1 \\ 3 \end{pmatrix} \quad \text{and} \quad \mathbf{r_2} = \begin{pmatrix} 2 \\ 9 \end{pmatrix} + t\begin{pmatrix} 2 \\ -1 \end{pmatrix}.$$

We want to find where the two lines meet, that is where $\mathbf{r_1} = \mathbf{r_2}$

$$\Rightarrow \quad \begin{pmatrix} 3 \\ -2 \end{pmatrix} + s\begin{pmatrix} 1 \\ 3 \end{pmatrix} = \begin{pmatrix} 2 \\ 9 \end{pmatrix} + t\begin{pmatrix} 2 \\ -1 \end{pmatrix}$$

$$\Rightarrow \quad \begin{pmatrix} 3 + s \\ -2 + 3s \end{pmatrix} = \begin{pmatrix} 2 + 2t \\ 9 - t \end{pmatrix}$$

$$\Rightarrow \quad \begin{cases} 3 + s = 2 + 2t \\ -2 + 3s = 9 - t \end{cases}.$$

Multiplying the second equation by 2 and then adding the two equations will produce an equation in s but not t:

$$
\begin{array}{rcrcr}
3 & + & s & = & 2 & + & 2t \\
\oplus \quad -4 & + & 6s & = & 18 & - & 2t \\
\hline
-1 & + & 7s & = & 20 & & \\
& \Rightarrow & 7s & = & 21 & & \\
& \Rightarrow & s & = & 3. & &
\end{array}
$$

The first equation is now $6 = 2 + 2t \Rightarrow t = 2$.

When $s = 3$, $\mathbf{r_1} = \begin{pmatrix} 3 \\ -2 \end{pmatrix} + 3\begin{pmatrix} 1 \\ 3 \end{pmatrix} = \begin{pmatrix} 6 \\ 7 \end{pmatrix}$

and when $t = 2$, $\mathbf{r_2} = \begin{pmatrix} 2 \\ 9 \end{pmatrix} + 2\begin{pmatrix} 2 \\ -1 \end{pmatrix} = \begin{pmatrix} 6 \\ 7 \end{pmatrix}$.

The two lines intersect at the point $(6, 7)$.

EXERCISE 2

1. Find vector equations for
   a) $y = 5x + 3$ b) $y = -3x + 2$ c) $x = 3$
   d) the line that passes through the points (2, 7) and (4, 10);
   e) the line that passes through the points (−1, 4) and (3, 5).

2. Find the Cartesian equation of each of the lines
   a) $\underline{r} = \begin{pmatrix} 3 \\ 4 \end{pmatrix} + t\begin{pmatrix} 1 \\ 5 \end{pmatrix}$ b) $\underline{r} = \begin{pmatrix} 3 \\ 5 \end{pmatrix} + t\begin{pmatrix} -2 \\ 1 \end{pmatrix}$

3. Find the points of intersection of the lines
   a) $\underline{r_1} = \begin{pmatrix} 3 \\ 5 \end{pmatrix} + s\begin{pmatrix} 2 \\ 1 \end{pmatrix}$ $\underline{r_2} = \begin{pmatrix} 2 \\ -9 \end{pmatrix} + t\begin{pmatrix} 1 \\ 5 \end{pmatrix}$

   b) $\underline{r_1} = s\begin{pmatrix} 3 \\ -2 \end{pmatrix}$ $\underline{r_2} = \begin{pmatrix} 3 \\ 4 \end{pmatrix} + t\begin{pmatrix} 1 \\ 1 \end{pmatrix}$

   c) $\underline{r_1} = \begin{pmatrix} 1 \\ 0 \end{pmatrix} + s\begin{pmatrix} 2 \\ -3 \end{pmatrix}$ $\underline{r_2} = \begin{pmatrix} 5 \\ -6 \end{pmatrix} + t\begin{pmatrix} -4 \\ 6 \end{pmatrix}$

   d) $\underline{r_1} = \begin{pmatrix} 3 \\ 2 \end{pmatrix} + s\begin{pmatrix} 1 \\ -1 \end{pmatrix}$ $\underline{r_2} = \begin{pmatrix} 5 \\ 1 \end{pmatrix} + t\begin{pmatrix} -2 \\ 2 \end{pmatrix}$

   e) $\underline{r_1} = \begin{pmatrix} 3 \\ 1 \end{pmatrix} + s\begin{pmatrix} 1 \\ 0 \end{pmatrix}$ $\underline{r_2} = \begin{pmatrix} 5 \\ 3 \end{pmatrix} + t\begin{pmatrix} 1 \\ -1 \end{pmatrix}$

   f) $\underline{r_1} = \begin{pmatrix} 7 \\ 11 \end{pmatrix} + s\begin{pmatrix} 2 \\ 1 \end{pmatrix}$ $\underline{r_2} = \begin{pmatrix} 0 \\ 3 \end{pmatrix} + t\begin{pmatrix} 2 \\ 4 \end{pmatrix}$

Vectors in Three Dimensions

Vectors are particularly useful when working with three-dimensional co-ordinate geometry.

Points in three dimensions are described by triplets (x, y, z) where the three axes are perpendicular to each other. It is usual to draw the axes so that the positive direction of the x-axis points to the right, the positive direction of the y-axis goes directly into the sheet of paper and the positive direction of the z-axis points directly up the page.

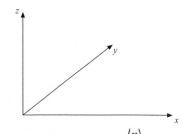

Movements in three dimensions are described by column vectors with three entries $\begin{pmatrix} p \\ q \\ r \end{pmatrix}$ where

p is the movement in the x-direction, q is the movement in the y-direction and r is the movement in the z-direction.

For example, If A(3, 2, −4) and B(6, −5, 2) then $\overrightarrow{AB} = \begin{pmatrix} 3 \\ -7 \\ 6 \end{pmatrix}$.

The length of three-dimensional vectors can be calculated using Pythagoras's theorem:

the length of $\begin{pmatrix} 1 \\ 3 \\ 4 \end{pmatrix}$ is the length of the line joining the origin to

R(1, 3, 4).

If K and L are the points (1, 0, 0) and (1, 3, 0), then Pythagoras's theorem applied to triangle OKL gives

$$OL^2 = OK^2 + KL^2$$
$$\Rightarrow \quad OL^2 = 1^2 + 3^2 = 10$$
$$\Rightarrow \quad OL = \sqrt{10}.$$

Applying Pythagoras's theorem to triangle OLR gives

$$OR^2 = OL^2 + LR^2$$
$$\Rightarrow \quad OR^2 = \left(\sqrt{10}\right)^2 + 4^2 = 26$$
$$\Rightarrow \quad OR = \sqrt{26} = \sqrt{1^2 + 3^2 + 4^2}.$$

The argument can be generalised to give the important result:

The length of $\begin{pmatrix} p \\ q \\ r \end{pmatrix}$ is given by $\sqrt{p^2 + q^2 + r^2}$.

The Vector Equation of a Line in Three Dimensions

We know that the equation $\underline{r} = \underline{a} + t\underline{b}$ represents a line which

- passes through the point A whose position vector is $\underline{a}$,
- has direction vector $\underline{b}$.

This result allows the vector equation of a line in three dimensions to be written quickly and easily.

EXAMPLE 7

Find a vector equation of the line passing through the points A and B whose co-ordinates are (1, 2, −3) and (3, 1, 2) respectively. Find also the co-ordinates of the point where the line meets the x–y plane.

SOLUTION

It is clear that $\overrightarrow{AC} = \begin{pmatrix} 2 \\ -1 \\ 5 \end{pmatrix}$ is a direction vector for the line.

The line passes through A(1, 2, −3) so a possible equation of the line is

$$\underline{r} = \begin{pmatrix} 1 \\ 2 \\ -3 \end{pmatrix} + t\begin{pmatrix} 2 \\ -1 \\ 5 \end{pmatrix}.$$

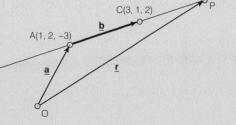

EXAMPLE 7 (continued)

The x–y plane is the plane (two-dimensional space) containing the x-axis and the y-axis. All the points on this plane have a z co-ordinate which is zero so the x–y plane can also be described as the plane of points where $z = 0$.

$$\mathbf{r} = \begin{pmatrix} 1 \\ 2 \\ -3 \end{pmatrix} + t \begin{pmatrix} 2 \\ -1 \\ 5 \end{pmatrix} \text{ and } z = 0$$

$$\Rightarrow \quad -3 + 5t = 0$$
$$\Rightarrow \quad t = 0.6$$
$$\Rightarrow \quad \mathbf{r} = \begin{pmatrix} 2.2 \\ 1.4 \\ 0 \end{pmatrix}.$$

So the line meets the x–y plane at the point $(2.2, 1.4, 0)$.

The **Cartesian equation of a line in three dimensions** is much more complicated than the vector equation.

For example, consider the line of the previous example whose vector equation was

$$\mathbf{r} = \begin{pmatrix} 1 \\ 2 \\ -3 \end{pmatrix} + t \begin{pmatrix} 2 \\ -1 \\ 5 \end{pmatrix}.$$

If the point P with co-ordinates (x, y, z) is on this line then

$$\begin{pmatrix} x \\ y \\ z \end{pmatrix} = \mathbf{r} = \begin{pmatrix} 1 \\ 2 \\ -3 \end{pmatrix} + t \begin{pmatrix} 2 \\ -1 \\ 5 \end{pmatrix} \Rightarrow \begin{cases} x = 1 + 2t \\ y = 2 - t \\ z = -3 + 5t \end{cases} \Rightarrow \frac{x - 1}{2} = \frac{y - 2}{-1} = \frac{z + 3}{5} \; (= t).$$

The Cartesian representation of the line is $\dfrac{x - 1}{2} = \dfrac{y - 2}{-1} = \dfrac{z + 3}{5}$.

In general, the line whose vector equation is $\mathbf{r} = \begin{pmatrix} a_1 \\ a_2 \\ a_3 \end{pmatrix} + t \begin{pmatrix} d_1 \\ d_2 \\ d_3 \end{pmatrix}$ has the Cartesian representation

$$\frac{x - a_1}{d_1} = \frac{y - a_2}{d_2} = \frac{z - a_3}{d_3}.$$

Notice that the numbers being subtracted in the numerator of the Cartesian representation are the co-ordinates of the fixed point on the line whilst the denominators are the components of the direction vector of the line.

The Intersection of Lines in Three Dimensions

Consider the cuboid shown in the diagram.

The lines RS and LR intersect at the point R.

The lines RL and SM will not intersect since they are parallel lines.

Now consider the lines KM and PR: the lines are certainly not parallel but nor do they intersect. The lines KM and PR are said to be skew lines.

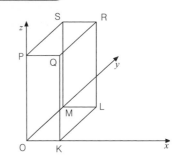

Skew lines are lines which are not parallel and which do not intersect.

It is easy to see whether two lines are parallel: for parallel lines the direction vectors must be parallel to each other. In other words, **the direction vectors of parallel lines must be multiples of each other**.

Given two non-parallel lines

$$\mathbf{\underline{r}_1} = \begin{pmatrix} a_1 \\ a_2 \\ a_3 \end{pmatrix} + s\begin{pmatrix} b_1 \\ b_2 \\ b_3 \end{pmatrix} \quad \text{and} \quad \mathbf{\underline{r}_2} = \begin{pmatrix} c_1 \\ c_2 \\ c_3 \end{pmatrix} + t\begin{pmatrix} d_1 \\ d_2 \\ d_3 \end{pmatrix}$$

the strategy used to determine whether they intersect or are skew is:

find the values of s and t so that two of the co-ordinates of each line are equal (for example $x_1 = x_2$ and $y_1 = y_2$), **and then for these values of s and t determine whether or not the third co-ordinates of each line are equal.**

If the third co-ordinates of each line are equal then the lines intersect.

If the third co-ordinates of each line are not equal then the lines are skew.

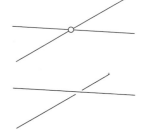

EXAMPLE 8

Determine whether the following lines intersect, are parallel or are skew.

$$L_1: \quad \mathbf{\underline{r}_1} = \begin{pmatrix} 1 \\ -2 \\ 0 \end{pmatrix} + s\begin{pmatrix} 1 \\ -2 \\ 1 \end{pmatrix} \qquad L_2: \quad \mathbf{\underline{r}_2} = \begin{pmatrix} -2 \\ -6 \\ -1 \end{pmatrix} + t\begin{pmatrix} 2 \\ 1 \\ 1 \end{pmatrix}$$

$$L_3: \quad \mathbf{\underline{r}_3} = \begin{pmatrix} 4 \\ 1 \\ 1 \end{pmatrix} + u\begin{pmatrix} -2 \\ 4 \\ -2 \end{pmatrix}$$

EXAMPLE 8 (continued)

The lines L_1 and L_3 **are parallel** because their direction vectors are parallel to each

other since $\begin{pmatrix} -2 \\ 4 \\ -2 \end{pmatrix} = -2 \begin{pmatrix} 1 \\ -2 \\ 1 \end{pmatrix}$.

> Always look at the direction vectors of the lines first to see whether they are parallel.
> Remember that if the direction vector of one line is a multiple of the direction vector of the other line then the lines must be parallel.

Now consider L_1 and L_2: they are certainly not parallel since there is no number k such that

$$\begin{pmatrix} 2 \\ 1 \\ 1 \end{pmatrix} = k \begin{pmatrix} 1 \\ -2 \\ 1 \end{pmatrix}.$$

To determine whether the lines intersect or are skew

- find the values of s and t so that $x_1 = x_2$ and $y_1 = y_2$,
- then for these values of s and t, if $z_1 = z_2$ then the lines intersect; if $z_1 \neq z_2$ then the lines are skew.

$$L_1: \mathbf{r_1} = \begin{pmatrix} 1 \\ -2 \\ 0 \end{pmatrix} + s \begin{pmatrix} 1 \\ -2 \\ 1 \end{pmatrix} \qquad L_2: \mathbf{r_2} = \begin{pmatrix} -2 \\ -6 \\ -1 \end{pmatrix} + t \begin{pmatrix} 2 \\ 1 \\ 1 \end{pmatrix}$$

$$\left. \begin{array}{c} x_1 = x_2 \\ y_1 = y_2 \end{array} \right\} \Rightarrow \left\{ \begin{array}{l} 1 + s = -2 + 2t \\ -2 - 2s = -6 + t \end{array} \right.$$

Multiplying the first equation by 2 and then adding

$$\Rightarrow \left\{ \begin{array}{l} 2 + 2s = -4 + 4t \\ -2 - 2s = -6 + t \end{array} \right. \Rightarrow 0 = -10 + 5t \Rightarrow t = 2.$$

Substituting for t in the first equation gives

$$1 + s = 2 \Rightarrow s = 1.$$

When $s = 1$ and $t = 2$,

$$z_1 = s = 1 \qquad z_2 = -1 + t = 1.$$

Since $z_1 = z_2$, the two lines **do intersect**.

When $s = 1$, $\mathbf{r_1} = \begin{pmatrix} 1 \\ -2 \\ 0 \end{pmatrix} + \begin{pmatrix} 1 \\ -2 \\ 1 \end{pmatrix} = \begin{pmatrix} 2 \\ -4 \\ 1 \end{pmatrix}$.

When $t = 2$, $\mathbf{r_2} = \begin{pmatrix} -2 \\ -6 \\ -1 \end{pmatrix} + 2 \begin{pmatrix} 2 \\ 1 \\ 1 \end{pmatrix} = \begin{pmatrix} 2 \\ -4 \\ 1 \end{pmatrix}$.

The lines L_1 and L_2 intersect at the point $(2, -4, 1)$.

Now consider the lines L_2 and L_3.

$$L_2: \mathbf{r_2} = \begin{pmatrix} -2 \\ -6 \\ -1 \end{pmatrix} + t \begin{pmatrix} 2 \\ 1 \\ 1 \end{pmatrix} \qquad L_3: \mathbf{r_3} = \begin{pmatrix} 4 \\ 1 \\ 1 \end{pmatrix} + u \begin{pmatrix} -2 \\ 4 \\ -2 \end{pmatrix}.$$

EXAMPLE 8 (continued)

The lines are not parallel since there is no number k such that $\begin{pmatrix} -2 \\ 4 \\ -2 \end{pmatrix} = k \begin{pmatrix} 2 \\ 1 \\ 1 \end{pmatrix}$.

To determine whether the lines intersect or are skew

- find the values of s and t so that $x_1 = x_2$ and $y_1 = y_2$,
- then for these values of s and t if $z_1 = z_2$ then the lines intersect; if $z_1 \neq z_2$ then the lines are skew.

$$\left. \begin{array}{c} x_2 = x_3 \\ y_2 = y_3 \end{array} \right\} \implies \begin{cases} -2 + 2t = 4 - 2u \\ -6 + t = 1 + 4u \end{cases}.$$

Multiplying the second equation by 2 and subtracting the result from the first

$$\implies \begin{cases} -2 + 2t = 4 - 2u \\ -12 + 2t = 2 + 8u \end{cases} \implies 10 = 2 - 10u \implies u = -\frac{4}{5}.$$

Substituting for u in the second equation gives

$$-6 + t = -\frac{11}{5} \implies t = \frac{19}{5}.$$

When $t = \frac{19}{5}$ and $u = \frac{-4}{5}$,

$$z_2 = -1 + t = \frac{14}{5} \qquad z_3 = 1 - 2u = \frac{13}{5}.$$

Notation

The notation used so far for writing three dimensional vectors is quick and easy when vectors are written by hand. However, in printed work, the notation can be awkward and consume large amounts of paper.

Vectors are therefore sometimes written in terms of the unit vectors (i.e. vectors of length 1 unit) $\underline{i}$, $\underline{j}$, $\underline{k}$ defined by

$$\underline{i} = \begin{pmatrix} 1 \\ 0 \\ 0 \end{pmatrix}, \quad \underline{j} = \begin{pmatrix} 0 \\ 1 \\ 0 \end{pmatrix}, \quad \underline{k} = \begin{pmatrix} 0 \\ 0 \\ 1 \end{pmatrix}.$$

Geometrically, the vector $\underline{i}$ is a movement of one unit in the direction of the positive x-axis; the vector $\underline{j}$ is a movement of one unit in the direction of the positive y-axis and the vector $\underline{k}$ is a movement of one unit in the direction of the positive z-axis.

Since $\begin{pmatrix} 4 \\ 3 \\ 5 \end{pmatrix} = \begin{pmatrix} 4 \\ 0 \\ 0 \end{pmatrix} + \begin{pmatrix} 0 \\ 3 \\ 0 \end{pmatrix} + \begin{pmatrix} 0 \\ 0 \\ 5 \end{pmatrix} = 4\begin{pmatrix} 1 \\ 0 \\ 0 \end{pmatrix} + 3\begin{pmatrix} 0 \\ 1 \\ 0 \end{pmatrix} + 5\begin{pmatrix} 0 \\ 0 \\ 1 \end{pmatrix}$, we can write the vector $\begin{pmatrix} 4 \\ 3 \\ 5 \end{pmatrix}$ as $4\underline{i} + 3\underline{j} + 5\underline{k}$.

Similarly, $7\underline{i} - 2\underline{j} - 3\underline{k}$ is a useful alternative way of writing the vector $\begin{pmatrix} 7 \\ -2 \\ -3 \end{pmatrix}$.

EXERCISE 3

1 Evaluate

a) $4\begin{pmatrix} 1 \\ 2 \\ -1 \end{pmatrix} - 2\begin{pmatrix} 0 \\ -1 \\ 3 \end{pmatrix}$ **b)** $3\begin{pmatrix} 4 \\ 1 \\ 2 \end{pmatrix} + 2\begin{pmatrix} 1 \\ -1 \\ 2 \end{pmatrix}$ **c)** $(5\underline{i} + 3\underline{j} - 2\underline{k}) + 2(3\underline{i} - \underline{j} + 2\underline{k})$

2 Find the length of the vectors

a) $\begin{pmatrix} 3 \\ 12 \\ -4 \end{pmatrix}$ **b)** $\begin{pmatrix} 3 \\ -2 \\ 2 \end{pmatrix}$ **c)** $5\underline{i} + 3\underline{j} - 2\underline{k}$

3 In each of the following, find a vector equation of the line passing through the given point:
a) $(0, 0, 2)$ and $(2, 1, 0)$ **b)** $(1, 5, 7)$ and $(3, 5, 7)$
c) $(1, 2, 4)$ and $(2, 4, 8)$

4 Find the Cartesian equation of each of the following lines:

a) $\underline{r} = \begin{pmatrix} 1 \\ 2 \\ 6 \end{pmatrix} + t\begin{pmatrix} 2 \\ 1 \\ 3 \end{pmatrix}$ **b)** $\underline{r} = \begin{pmatrix} 3 \\ 0 \\ 6 \end{pmatrix} + t\begin{pmatrix} -4 \\ 2 \\ 5 \end{pmatrix}$

5 Determine whether the following pairs of lines are parallel, intersecting or skew. In the case of intersecting lines, find the point of intersection:

a) $\underline{r}_1 = \begin{pmatrix} 1 \\ -1 \\ 1 \end{pmatrix} + s\begin{pmatrix} 3 \\ -4 \\ 1 \end{pmatrix}$ $\underline{r}_2 = t\begin{pmatrix} -9 \\ 12 \\ -3 \end{pmatrix}$

b) $\underline{r}_1 = 4\underline{i} + 8\underline{j} + 3\underline{k} + s(\underline{i} + 2\underline{j} + \underline{k})$ $\underline{r}_2 = 7\underline{i} + 6\underline{j} + 5\underline{k} + t(6\underline{i} + 4\underline{j} + 5\underline{k})$

c) $\underline{r}_1 = \begin{pmatrix} 1 \\ 0 \\ 3 \end{pmatrix} + s\begin{pmatrix} 2 \\ 1 \\ 1 \end{pmatrix}$ $\underline{r}_2 = \begin{pmatrix} 2 \\ -1 \\ 1 \end{pmatrix} + t\begin{pmatrix} 1 \\ -2 \\ 0 \end{pmatrix}$

6 Two intersecting lines have equations

$$\underline{r}_1 = \begin{pmatrix} 2 \\ 9 \\ 13 \end{pmatrix} + s\begin{pmatrix} 1 \\ 2 \\ 3 \end{pmatrix} \qquad \underline{r}_2 = \begin{pmatrix} \lambda \\ 7 \\ -2 \end{pmatrix} + t\begin{pmatrix} -1 \\ 2 \\ -3 \end{pmatrix}.$$

Find the value of λ and the co-ordinates of the point of intersection.

7 Show that the lines

$\underline{r}_1 = 2\underline{i} - \underline{j} + \underline{k} + s(\underline{i} - 2\underline{j} + 2\underline{k})$

$\underline{r}_2 = \underline{i} - 3\underline{j} + 4\underline{k} + t(2\underline{i} + 3\underline{j} - 6\underline{k})$

are skew.

The Scalar Product (or Dot Product) of Two Vectors

We have now established ways of describing lines in three-dimensional space and ways of determining whether or not the lines intersect. The next aim is to establish an arithmetic procedure for determining angles which can be applied easily in both two- and three-dimensional problems. The key idea is the **scalar product** of two vectors which is defined below.

Definition
The scalar product of two vectors, $\underline{a}$ and $\underline{b}$, is written
$\underline{a} \cdot \underline{b}$ and is defined by

$$\underline{a} \cdot \underline{b} = ab \cos \theta$$

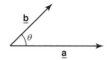

where a and b are the lengths of the vectors $\underline{a}$ and $\underline{b}$ respectively and θ is the angle between the two vectors.

For example, for the vectors shown in the diagram
$\underline{a} \cdot \underline{b} = 4 \times 5 \times \cos 60° = 10$
$\underline{b} \cdot \underline{c} = 5 \times 3 \times \cos 90° = 0$
$\underline{c} \cdot \underline{a} = 3 \times 4 \times \cos 150° = -10.39$ (2 d.p.)

Properties of the Scalar Product

If $\underline{a}$ and $\underline{b}$ are perpendicular then the angle between the vectors $\underline{a}$ and $\underline{b}$ is 90° so

$$\underline{a} \cdot \underline{b} = ab \cos 90° = 0.$$

Conversely, if $\underline{a} \cdot \underline{b} = 0$, then either $a = 0$ or $b = 0$ or $\cos \theta = 0$.

Thus if $\underline{a}$ and $\underline{b}$ are two non-zero vectors with $\underline{a} \cdot \underline{b} = 0$ then $\underline{a}$ and $\underline{b}$ must be perpendicular.

> **The two non-zero vectors $\underline{a}$ and $\underline{b}$ are perpendicular if and only if $\underline{a} \cdot \underline{b} = 0$**

If we have two column vectors $\underline{a} = \begin{pmatrix} a_1 \\ a_2 \\ a_3 \end{pmatrix}$ and $\underline{b} = \begin{pmatrix} b_1 \\ b_2 \\ b_3 \end{pmatrix}$ then the cosine rule for triangles and some algebraic manipulation give a quick and easy way of evaluating $\underline{a} \cdot \underline{b}$:

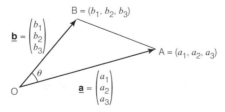

Let O be the origin, A be the point (a_1, a_2, a_3) and B be the point (b_1, b_2, b_3).

The cosine rule applied to triangle OAB gives

$$AB^2 = OA^2 + OB^2 - 2 \times OA \times OB \times \cos \theta$$
$$\Rightarrow \quad AB^2 = OA^2 + OB^2 - 2 \times \underline{a} \cdot \underline{b}$$
$$\Rightarrow \quad \underline{a} \cdot \underline{b} = \frac{OA^2 + OB^2 - AB^2}{2}.$$

> OA is the length of vector $\underline{a}$, OB is the length of vector $\underline{b}$ so
> $OA \times OB \cos \theta = \underline{a} \cdot \underline{b}$.

Now

$$\text{OA} = \text{length of} \begin{pmatrix} a_1 \\ a_2 \\ a_3 \end{pmatrix} = \sqrt{a_1^2 + a_2^2 + a_3^2} \, , \ \text{OB} = \text{length of} \begin{pmatrix} b_1 \\ b_2 \\ b_3 \end{pmatrix} = \sqrt{b_1^2 + b_2^2 + b_3^2}$$

and

$$\text{AB} = \text{length of} \ \overrightarrow{\textbf{AB}} = \text{length of} \begin{pmatrix} b_1 - a_1 \\ b_2 - a_2 \\ b_3 - a_3 \end{pmatrix} = \sqrt{(b_1 - a_1)^2 + (b_2 - a_2)^2 + (b_3 - a_3)^2}.$$

These results mean that

$$\underline{\textbf{a}} \cdot \underline{\textbf{b}} = \frac{(a_1^2 + a_2^2 + a_3^2) + (b_1^2 + b_2^2 + b_3^2) - ((b_1 - a_1)^2 + (b_2 - a_2)^2 + (b_3 - a_3)^2)}{2}$$

$$= \frac{(a_1^2 + a_2^2 + a_3^2) + (b_1^2 + b_2^2 + b_3^2) - (b_1^2 - 2a_1 b_1 + a_1^2 + b_2^2 - 2a_2 b_2 + a_2^2 + b_3^2 - 2a_3 b_3 + a_3^2)}{2}$$

$$= \frac{2a_1 b_1 + 2a_2 b_2 + 2a_3 b_3}{2}$$

$$= a_1 b_1 + a_2 b_2 + a_3 b_3.$$

We have demonstrated that if $\underline{\textbf{a}} = \begin{pmatrix} a_1 \\ a_2 \\ a_3 \end{pmatrix}$ and $\underline{\textbf{b}} = \begin{pmatrix} b_1 \\ b_2 \\ b_3 \end{pmatrix}$ then $\underline{\textbf{a}} \cdot \underline{\textbf{b}} = a_1 b_1 + a_2 b_2 + a_3 b_3$.

The result for two-dimensional vectors can be deduced from this result by merely putting $a_3 = b_3 = 0$ to obtain

if $\quad \underline{\textbf{a}} = \begin{pmatrix} a_1 \\ a_2 \end{pmatrix} \quad$ and $\quad \underline{\textbf{b}} = \begin{pmatrix} b_1 \\ b_2 \end{pmatrix} \quad$ then $\quad \underline{\textbf{a}} \cdot \underline{\textbf{b}} = a_1 b_1 + a_2 b_2.$

We therefore have

$$\begin{pmatrix} a_1 \\ a_2 \\ a_3 \end{pmatrix} \cdot \begin{pmatrix} b_1 \\ b_2 \\ b_3 \end{pmatrix} = a_1 b_1 + a_2 b_2 + a_3 b_3 \qquad \begin{pmatrix} a_1 \\ a_2 \end{pmatrix} \cdot \begin{pmatrix} b_1 \\ b_2 \end{pmatrix} = a_1 b_1 + a_2 b_2$$

EXAMPLE 9

Evaluate

a) $\begin{pmatrix} 4 \\ -1 \end{pmatrix} \cdot \begin{pmatrix} 2 \\ 3 \end{pmatrix}$

b) $(\underline{\textbf{i}} + 3\underline{\textbf{j}} - 4\underline{\textbf{k}}) \cdot (2\underline{\textbf{i}} - 3\underline{\textbf{j}} - 3\underline{\textbf{k}})$

EXAMPLE 9 (continued)

a) $\begin{pmatrix} 4 \\ -1 \end{pmatrix} \cdot \begin{pmatrix} 2 \\ 3 \end{pmatrix} = 4 \times 2 + (-1) \times 3 = 8 - 3 = 5$

b) $(\underline{i} + 3\underline{j} - 4\underline{k}) \cdot (2\underline{i} - 3\underline{j} - 3\underline{k}) = \begin{pmatrix} 1 \\ 3 \\ -4 \end{pmatrix} \cdot \begin{pmatrix} 2 \\ -3 \\ -3 \end{pmatrix}$

$$= 1 \times 2 + 3 \times (-3) + (-4) \times (-3)$$
$$= 2 + (-9) + 12$$
$$= 5.$$

EXAMPLE 10

The vectors $\begin{pmatrix} 3 \\ 2 \\ -5 \end{pmatrix}$ and $\begin{pmatrix} 4 \\ -1 \\ k \end{pmatrix}$ are perpendicular. Find the value of k.

Since the vectors are perpendicular, their scalar product must be 0:

$$\begin{pmatrix} 3 \\ 2 \\ -5 \end{pmatrix} \cdot \begin{pmatrix} 4 \\ -1 \\ k \end{pmatrix} = 0$$

$\Longrightarrow \quad 12 + (-2) + (-5k) = 0$

$\Longrightarrow \quad 10 - 5k = 0$

$\Longrightarrow \quad k = 2.$

EXERCISE 4

1 The diagram shows a system of four coplanar vectors.
(This means that the four vectors all lie in the same plane.)
Find the values of

a) $\underline{p} \cdot \underline{q}$

b) $\underline{p} \cdot \underline{r}$

c) $\underline{p} \cdot \underline{s}$

d) $\underline{p} \cdot \underline{p}$

e) $\underline{q} \cdot \underline{r}$

f) $\underline{q} \cdot \underline{s}$

g) $\underline{r} \cdot \underline{s}$

h) $\underline{s} \cdot \underline{s}$.

2 Evaluate

a) $\begin{pmatrix} 4 \\ 3 \end{pmatrix} \cdot \begin{pmatrix} -1 \\ 2 \end{pmatrix}$

b) $\begin{pmatrix} -3 \\ -4 \end{pmatrix} \cdot \begin{pmatrix} 3 \\ -3 \end{pmatrix}$

c) $\begin{pmatrix} 2 \\ 5 \end{pmatrix} \cdot \begin{pmatrix} 1 \\ -4 \end{pmatrix}$

d) $\begin{pmatrix} 3 \\ 1 \\ -5 \end{pmatrix} \cdot \begin{pmatrix} 1 \\ -2 \\ -2 \end{pmatrix}$

e) $\begin{pmatrix} 1 \\ 0 \\ -1 \end{pmatrix} \cdot \begin{pmatrix} 3 \\ 5 \\ 3 \end{pmatrix}$

f) $3\underline{j} \cdot (2\underline{i} - 4\underline{j} + 2\underline{k})$

3 If $\underline{a} = \begin{pmatrix} 4 \\ -6 \\ 1 \end{pmatrix}$, $\underline{b} = \begin{pmatrix} 7 \\ 5 \\ 2 \end{pmatrix}$ evaluate $\underline{a} \cdot \underline{b}$. What can be deduced about vectors $\underline{a}$ and $\underline{b}$?

4 i) The vectors $\underline{a} = \begin{pmatrix} 2 \\ -2 \\ 3 \end{pmatrix}$ and $\underline{b} = \begin{pmatrix} 4 \\ 1 \\ p \end{pmatrix}$ are perpendicular.

Find the value of p.

ii) The vector $\underline{c} = \begin{pmatrix} u \\ v \\ 10 \end{pmatrix}$ is perpendicular to both $\underline{a}$ and $\underline{b}$. Find the values of u and v.

5 If $\underline{a} = \begin{pmatrix} a_1 \\ a_2 \\ a_3 \end{pmatrix}$, $\underline{b} = \begin{pmatrix} b_1 \\ b_2 \\ b_3 \end{pmatrix}$ and $\underline{c} = \begin{pmatrix} c_1 \\ c_2 \\ c_3 \end{pmatrix}$

prove that $\underline{a} \cdot (\underline{b} + \underline{c}) = \underline{a} \cdot \underline{b} + \underline{a} \cdot \underline{c}$.

Using the Scalar Product to Find the Angle between Two Vectors or the Angle between Two Lines

Putting the definition of the scalar product together with the result giving the method of evaluation of the scalar product of two column vectors gives

$$a_1 b_1 + a_2 b_2 + a_3 b_3 = \begin{pmatrix} a_1 \\ a_2 \\ a_3 \end{pmatrix} \cdot \begin{pmatrix} b_1 \\ b_2 \\ b_3 \end{pmatrix} = \underline{a} \cdot \underline{b} = ab \cos \theta$$

and this can be used to determine the angle between two vectors.

EXAMPLE 11

Determine, correct to one decimal place, the angle, in degrees, between the vectors $\begin{pmatrix} 1 \\ 5 \\ -2 \end{pmatrix}$ and $\begin{pmatrix} 3 \\ -2 \\ -1 \end{pmatrix}$.

We know that we can write

$$\begin{pmatrix} 1 \\ 5 \\ -2 \end{pmatrix} \cdot \begin{pmatrix} 3 \\ -2 \\ -1 \end{pmatrix} = 3 + (-10) + 2 = -5. \qquad [1]$$

If θ is the angle between the two vectors then the initial definition of the scalar product implies that

$$\begin{pmatrix} 1 \\ 5 \\ -2 \end{pmatrix} \cdot \begin{pmatrix} 3 \\ -2 \\ -1 \end{pmatrix} = \sqrt{30}\sqrt{14} \cos \theta. \qquad [2]$$

Length of $\begin{pmatrix} 1 \\ 5 \\ -2 \end{pmatrix}$ is $\sqrt{1^2 + 5^2 + (-2)^2} = \sqrt{30}$.

Length of $\begin{pmatrix} 3 \\ -2 \\ -1 \end{pmatrix}$ is $\sqrt{3^2 + (-2)^2 + (-1)^2} = \sqrt{14}$.

EXAMPLE 11 (continued)

Combining [1] and [2] gives

$$-5 = \sqrt{30}\sqrt{14} \cos \theta$$

$$\Rightarrow \quad \cos \theta = -\frac{5}{\sqrt{30}\sqrt{14}}$$

$$\Rightarrow \quad \theta = 104.1° \quad (1 \text{ d.p.})$$

The angle between the two vectors is 104.1°, correct to one decimal place.

EXAMPLE 12

Find the angle between the two lines

$$L_1: \underline{r_1} = \begin{pmatrix} 1 \\ -2 \\ 0 \end{pmatrix} + s \begin{pmatrix} 1 \\ -2 \\ 1 \end{pmatrix} \qquad L_2: \underline{r_2} = \begin{pmatrix} -2 \\ -6 \\ -1 \end{pmatrix} + t \begin{pmatrix} 2 \\ 1 \\ 1 \end{pmatrix}.$$

We already know from example 8 that the two lines intersect at the point (2, −4, 1).

It is helpful to illustrate the problem with a simplified diagram.

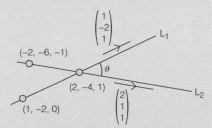

From the diagram it can be seen that the angle between the two lines is precisely the same as **the angle between the two direction vectors**.

The angle between the two lines is therefore the angle between the vectors

$$\begin{pmatrix} 1 \\ -2 \\ 1 \end{pmatrix} \text{ and } \begin{pmatrix} 2 \\ 1 \\ 1 \end{pmatrix}.$$

If θ is the angle between these two vectors we know that

$$\begin{pmatrix} 1 \\ -2 \\ 1 \end{pmatrix} \cdot \begin{pmatrix} 2 \\ 1 \\ 1 \end{pmatrix} = 2 + (-2) + 1 = 1 \quad \text{and that} \quad \begin{pmatrix} 1 \\ -2 \\ 1 \end{pmatrix} \cdot \begin{pmatrix} 2 \\ 1 \\ 1 \end{pmatrix} = \sqrt{6}\sqrt{6} \cos \theta.$$

Combining the two results gives

$$1 = 6 \cos \theta$$

$$\Rightarrow \quad \cos \theta = \frac{1}{6}$$

$$\Rightarrow \quad \theta = 80.4° \quad (1 \text{ d.p.})$$

> Length of $\begin{pmatrix} 1 \\ -2 \\ 1 \end{pmatrix}$ is $\sqrt{1^2 + (-2)^2 + 1^2} = \sqrt{6}$.
>
> Length of $\begin{pmatrix} 2 \\ 1 \\ 1 \end{pmatrix}$ is $\sqrt{2^2 + 1^2 + 1^2} = \sqrt{6}$.

The angle between the two lines is 80.4°, correct to one decimal place.

EXAMPLE 13

A line has equation $\underline{r} = \begin{pmatrix} 2 \\ 0 \\ -1 \end{pmatrix} + t \begin{pmatrix} 2 \\ -2 \\ 1 \end{pmatrix}$.

a) Find the angle that the line makes with the positive x-direction.
b) Find the distance of the point A(0, −7, 7) from the line.
c) Find the image of the point A(0, −7, 7) after a reflection in the line.

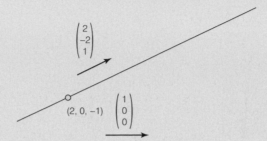

a) The angle between the line and the positive x-direction is the same as the angle between

$\begin{pmatrix} 2 \\ -2 \\ 1 \end{pmatrix}$ and $\begin{pmatrix} 1 \\ 0 \\ 0 \end{pmatrix}$.

If θ is the angle between these two vectors then evaluation of the scalar product of the two vectors in the two different ways gives

$$\begin{pmatrix} 2 \\ -2 \\ 1 \end{pmatrix} \cdot \begin{pmatrix} 1 \\ 0 \\ 0 \end{pmatrix} = 2 + 0 + 0 = 2 \quad \text{and} \quad \begin{pmatrix} 2 \\ -2 \\ 1 \end{pmatrix} \cdot \begin{pmatrix} 1 \\ 0 \\ 0 \end{pmatrix} = 3 \times 1 \times \cos \theta$$

Combining the two results gives

$2 = 3 \cos \theta$

$\Rightarrow \quad \cos \theta = \dfrac{2}{3}$

$\Rightarrow \quad \theta = 48.2°$ (1 d.p.)

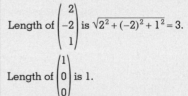

Length of $\begin{pmatrix} 2 \\ -2 \\ 1 \end{pmatrix}$ is $\sqrt{2^2 + (-2)^2 + 1^2} = 3$.

Length of $\begin{pmatrix} 1 \\ 0 \\ 0 \end{pmatrix}$ is 1.

The angle between the line and the positive x-direction is 48.2°, correct to one decimal place.

b) Now let P be the point on the line $\underline{r} = \begin{pmatrix} 2 \\ 0 \\ -1 \end{pmatrix} + t \begin{pmatrix} 2 \\ -2 \\ 1 \end{pmatrix}$ that is closest to the point A.

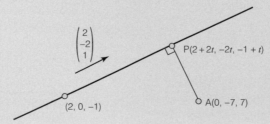

EXAMPLE 13 (continued)

Since P is on the line its co-ordinates must be of the form $(2 + 2t, -2t, -1 + t)$ for some value of t.

AP must be perpendicular to the line: in other words the vectors $\overrightarrow{AP}$ and $\begin{pmatrix} 2 \\ -2 \\ 1 \end{pmatrix}$ must be perpendicular. The scalar product of these two vectors therefore must be zero.

$$\overrightarrow{AP} = \begin{pmatrix} 2 + 2t \\ 7 - 2t \\ t - 8 \end{pmatrix}$$

$$\overrightarrow{AP} \cdot \begin{pmatrix} 2 \\ -2 \\ 1 \end{pmatrix} = 0 \quad \Longrightarrow \quad \begin{pmatrix} 2 + 2t \\ 7 - 2t \\ t - 8 \end{pmatrix} \cdot \begin{pmatrix} 2 \\ -2 \\ 1 \end{pmatrix} = 0$$

$$\Longrightarrow \quad 2(2 + 2t) + -2(7 - 2t) + 1(t - 8) = 0$$
$$\Longrightarrow \quad 9t - 18 = 0$$
$$\Longrightarrow \quad t = 2$$

$$\Longrightarrow \quad \overrightarrow{AP} = \begin{pmatrix} 6 \\ 3 \\ -6 \end{pmatrix}$$

$$\Longrightarrow \quad AP = \sqrt{6^2 + 3^2 + (-6)^2} = 9.$$

c) If A' is the image of A after the reflection then

$$\overrightarrow{AA'} = 2 \times \overrightarrow{AP} = 2\begin{pmatrix} 6 \\ 3 \\ -6 \end{pmatrix} = \begin{pmatrix} 12 \\ 6 \\ -12 \end{pmatrix}.$$

Since A(0, -7, 7), A' will be the point (12, -1, -5).

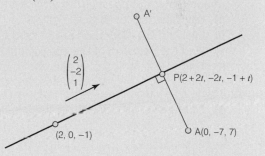

Angle between Skew Lines

We have seen that the angle between two intersecting lines is found by calculating the angle between the direction vectors of two lines.

The angle between two skew lines is **defined** to be the angle between the direction vectors of two lines.

EXERCISE 5

1 For each of the following pairs of vectors:
  a) calculate the length of each of the two vectors;
  b) calculate the scalar product of the two vectors;
  c) find the angle between the two vectors.

  i) $\begin{pmatrix} 3 \\ -4 \end{pmatrix}$ and $\begin{pmatrix} 1 \\ 1 \end{pmatrix}$ **ii)** $\begin{pmatrix} 7 \\ 8 \end{pmatrix}$ and $\begin{pmatrix} 16 \\ -14 \end{pmatrix}$ **iii)** $\begin{pmatrix} 3 \\ 1 \\ 2 \end{pmatrix}$ and $\begin{pmatrix} 2 \\ -2 \\ 1 \end{pmatrix}$

2 Find the angle between the following lines:

  a) $\underline{\mathbf{r}}_1 = s\begin{pmatrix} 1 \\ 0 \\ -3 \end{pmatrix}$ and $\underline{\mathbf{r}}_2 = \begin{pmatrix} 3 \\ 7 \\ 5 \end{pmatrix} + t\begin{pmatrix} 3 \\ 4 \\ 1 \end{pmatrix}$.

  b) $\underline{\mathbf{r}}_1 = s\begin{pmatrix} 1 \\ 0 \\ -3 \end{pmatrix}$ and $\underline{\mathbf{r}}_2 = \begin{pmatrix} 3 \\ 7 \\ 5 \end{pmatrix} + t\begin{pmatrix} 2 \\ 0 \\ -3 \end{pmatrix}$.

3 *In this question the point O represents an observation point, the x-axis points due East, the y-axis points due North and the z-axis points vertically upwards. Throughout the question all units of length are kilometres and all units of time are minutes.*

  A plane has taken off from an airfield which is on the same level as the observation point and flies in a straight line. The plane is initially spotted at the point A(4, 2, 5) and two minutes later disappears from view at the point B(−2, 7, 6).
  a) Calculate the speed of the plane.
  b) Evaluate $\overrightarrow{\mathbf{OA}} \cdot \overrightarrow{\mathbf{OB}}$ and hence find the size of ∠AOB.
  c) Find the equation of the line joining A and B.
  d) Find the co-ordinates of the take off point of the plane.

4 The line L_1 passes through the points A(1, 5, 3) and B(0, 7, 4).
  a) Find the vector equation of the line L_1.

  The line L_2 has equation $\underline{\mathbf{r}} = \begin{pmatrix} -4 \\ 8 \\ \lambda \end{pmatrix} + t\begin{pmatrix} 2 \\ 3 \\ -2 \end{pmatrix}$.

  The lines L_1 and L_2 are known to intersect at the point C.
  b) Find the value of λ and the co-ordinates of the point C.
  c) Find the acute angle between the two lines.
  d) What is the perpendicular distance of the point A from the line L_2?

5 The triangle ABC has vertices A(3, 1, 5), B(4, 2, 1) and C(3, 2, 7). Calculate
  a) the length of AB and AC;
  b) $\overrightarrow{\mathbf{AB}} \cdot \overrightarrow{\mathbf{AC}}$;
  c) the cosine of angle BAC, leaving your answer in surd form;
  d) the sine of angle BAC, leaving your answer in surd form.

  Hence prove that the area of the triangle ABC is $\dfrac{1}{2}\sqrt{41}$.

6 Given the points A(4, 2, 6) and B(7, 8, 9), write down a vector equation of the line AB. The perpendicular to the line AB from the point C(1, 8, 3) meets the line at N. Find the co-ordinates of N.
Obtain a vector equation for the line which is the reflection of the line AC in the line AB.

7 The lines L_1 and L_2 are given by the vector equations

$$\mathbf{r}_1 = \begin{pmatrix} 0 \\ 1 \\ -1 \end{pmatrix} + s \begin{pmatrix} 2 \\ -1 \\ 2 \end{pmatrix} \quad \text{and} \quad \mathbf{r}_2 = \begin{pmatrix} p \\ 3 \\ 0 \end{pmatrix} + t \begin{pmatrix} 2 \\ 2 \\ -1 \end{pmatrix}$$

respectively, where s and t are parameters. Given that the two lines intersect find the value of p. Find also the angle between these two lines.
Find the distance of the point $(p, 3, 0)$ from the point of intersection of the two lines.

8 The lines L_1 and L_2 are given by the vector equations

$$\mathbf{r}_1 = \begin{pmatrix} 1 \\ 6 \\ 3 \end{pmatrix} + s \begin{pmatrix} 2 \\ -1 \\ 1 \end{pmatrix} \quad \text{and} \quad \mathbf{r}_2 = \begin{pmatrix} 3 \\ 3 \\ 8 \end{pmatrix} + t \begin{pmatrix} 1 \\ 0 \\ 1 \end{pmatrix}.$$

a) Calculate the acute angle between the two lines.
b) Show that the lines do not intersect.

c) Show that the vector $\mathbf{a} = \begin{pmatrix} 1 \\ 1 \\ -1 \end{pmatrix}$ is perpendicular to each of these lines.

d) The point P on L_1 is given by $s = p$ so that P$(1 + 2p, 6 - p, 3 + p)$ and the point Q on L_2 is given by $t = q$.

Write down the column vector $\overrightarrow{PQ}$. Hence find the values of p and q so that $\overrightarrow{PQ}$ and $\mathbf{a}$ are parallel.

9 If OABC is a rhombus with $\overrightarrow{OA} = \mathbf{a}$ and $\overrightarrow{OC} = \mathbf{c}$ what can be said about the lengths of $\mathbf{a}$ and $\mathbf{c}$? Express $\overrightarrow{OB}$ and $\overrightarrow{AC}$ in terms of $\mathbf{a}$ and n terms of $\mathbf{a}$ and $\mathbf{c}$. Find $\overrightarrow{OB} \cdot \overrightarrow{AC}$. What does this tell you about the diagonals of a rhombus?

Having studied this chapter you should know

- how to add and subtract column vectors and multiply a column vector by a scalar and be able to interpret these operations geometrically

- that the vector $\mathbf{a} = \begin{pmatrix} a_1 \\ a_2 \\ a_3 \end{pmatrix}$ has length $\sqrt{a_1^2 + a_2^2 + a_3^2}$

- that the equation

  $$\mathbf{r} = \mathbf{a} + t\mathbf{b}$$

  gives the position vector $\mathbf{r}$ of an arbitrary point on the line with direction vector $\mathbf{b}$ passing through a point A whose position vector is $\mathbf{a}$

- how to determine whether two lines are parallel by examining whether the direction vector of the second line is a multiple of the direction vector of the first line

- how to determine whether two non-parallel lines are intersecting or skew

- the definition of the scalar product as

  $$\mathbf{a} \cdot \mathbf{b} = ab \cos \theta$$

  where a is the length of $\mathbf{a}$, b is the length of $\mathbf{b}$ and θ is the angle between the two vectors and be able to evaluate the scalar product of two vectors using the results.

  $$\begin{pmatrix} a_1 \\ a_2 \\ a_3 \end{pmatrix} \cdot \begin{pmatrix} b_1 \\ b_2 \\ b_3 \end{pmatrix} = a_1 b_1 + a_2 b_2 + a_3 b_3 \qquad \begin{pmatrix} a_1 \\ a_2 \end{pmatrix} \cdot \begin{pmatrix} b_1 \\ b_2 \end{pmatrix} = a_1 b_1 + a_2 b_2$$

- that two non-zero vectors are perpendicular if and only if $\mathbf{a} \cdot \mathbf{b} = 0$

- how to use the scalar product to find the angle between two vectors

- that the angle between two lines is found by calculating the angle between the direction vectors of the lines

REVISION EXERCISE

1 The vectors $\mathbf{p}$ and $\mathbf{q}$ are defined by $\mathbf{p} = \begin{pmatrix} 3 \\ 1 \\ -2 \end{pmatrix}$, $\mathbf{q} = \begin{pmatrix} 2 \\ -3 \\ 4 \end{pmatrix}$. The point A has position vector $\mathbf{a}$ given by $\mathbf{a} = 2\mathbf{p} - 3\mathbf{q}$.

   i)   Calculate the vector $\mathbf{a}$.
   ii) Calculate the length of the line OA where O is the origin.
   iii) Calculate $\mathbf{a} \cdot \mathbf{p}$.
   iv) Find the angle, correct to the nearest degree, between the vectors $\mathbf{a}$ and $\mathbf{p}$.

2 a) Write down a vector equation of the line, L, joining the points A(6, 3, 0) and B(8, 7, 4).

The line M has equation $\mathbf{r} = \begin{pmatrix} -1 \\ -3 \\ -8 \end{pmatrix} + s\begin{pmatrix} 3 \\ 2 \\ 3 \end{pmatrix}$.

b) Prove that the lines L and M intersect and find the co-ordinates of the point of intersection.

c) By using an appropriate scalar product, or otherwise, find the acute angle between the two lines, giving your answer to the nearest tenth of a degree.

3 The line L_1 passes through the point (3, 6, 1) and is parallel to the vector $2\mathbf{i} + 3\mathbf{j} - \mathbf{k}$. The line L_2 passes through (3, −1, 4) and is parallel to the vector $\mathbf{i} - 2\mathbf{j} + \mathbf{k}$.

i) Write down vector equations for the lines L_1 and L_2.

ii) Prove that L_1 and L_2 intersect and find the co-ordinates of the point of intersection.

iii) Calculate the acute angle between the lines.

(OCR Jun 2002 P3)

4 The vectors **a** and **b** have lengths 5 units and 3 units respectively. The angle between the directions of the two vectors is 120°.

i) Write down the value of $\mathbf{a} \cdot \mathbf{b}$.

ii) Draw a diagram showing the vectors **a**, **b** and 3**a** − 2**b**. Calculate the length of the vector 3**a** − 2**b**, giving the answer as a surd.

5 The foot F of a vertical radio mast is situated on horizontal ground. Relative to axes Oxyz, with Oz vertically upwards, the co-ordinates of F are (2, −8, 0), where the units are metres.

Two straight steel cables are attached to the mast at a point T, 20 m vertically above F. The other ends of the cables are attached to the ground at the points G(8, −6, 0) and H(−2, 4, 0).

i) Write down the co-ordinates of T.

ii) Write down the vector $\overrightarrow{GT}$ and deduce the length of the cable linking G and T.

iii) Calculate the value of the scalar product $\overrightarrow{GT} \cdot \overrightarrow{HT}$.

iv) Calculate, to the nearest degree, the angle between the two cables.

6 Two points A and B have position vectors $3\mathbf{i} - \mathbf{j} + 2\mathbf{k}$ and $2\mathbf{j} + 3\mathbf{k}$ respectively.

i) Find the vector $\overrightarrow{AB}$, and hence find the length of AB.

ii) Find a vector equation for the line through A and B.

iii) Show that the line through A and B does not intersect the line through the origin parallel to the vector **i**.

(OCR Jan 2004 P3)

7 The diagram shows a cube OABCDEFG with sides of length 2 units. Unit vectors **i**, **j**, **k** are directed along OA, OC and OD respectively. The midpoint of AB is M and the midpoint of CG is N.

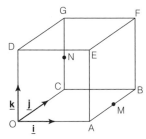

The point P is on the line MN such that $\overrightarrow{MP} = 2\overrightarrow{PN}$.

i) Verify that the position vector of P is $\frac{1}{3}(2\underline{i} + 5\underline{j} + 2\underline{k})$.

ii) Calculate the acute angle between the lines OP and MN.

iii) Show that the lines OP and EF intersect, and find the position vector of the point of intersection.

(OCR Jun 2004 P3)

8 A line L has equation $\underline{r} = \begin{pmatrix} 1 \\ 5 \\ 3 \end{pmatrix} + s\begin{pmatrix} 2 \\ 1 \\ -2 \end{pmatrix}$ where s is a parameter.

The point A has position vector $\begin{pmatrix} 9 \\ 8 \\ -1 \end{pmatrix}$ and the point P is the point on the line L with parameter p.

i) Show that $\overrightarrow{AP} = \begin{pmatrix} -8 + 2p \\ -3 + p \\ 4 - 2p \end{pmatrix}$.

ii) Determine the value of p if AP is perpendicular to the line L.

iii) Hence determine the perpendicular distance of A from the line L.

9 The position vectors of A and B with respect to the origin O are $\begin{pmatrix} 6 \\ 2 \\ 2 \end{pmatrix}$ and $\begin{pmatrix} 2 \\ 4 \\ 3 \end{pmatrix}$ respectively.

The points C and D are such that $\overrightarrow{OC} = \frac{3}{2}\overrightarrow{OA}$ and $\overrightarrow{OD} = 2\overrightarrow{OB}$.

i) State the position vectors of C and D.

ii) The line through A and D is denoted by L_1 and the line through B and C is denoted by L_2. Show that L_1 has equation

$$\underline{r} = \begin{pmatrix} 6 \\ 2 \\ 2 \end{pmatrix} + \lambda\begin{pmatrix} -2 \\ 6 \\ 4 \end{pmatrix}$$

and find an equation for L_2 in a similar form.

iii) The lines L_1 and L_2 intersect at X. Find the position vector of X.

iv) Calculate the acute angle between L_1 and L_2, correct to the nearest degree.

(OCR Jun 2001 P3)

10 In this question, the x-axis and the y-axis lie on a horizontal plane which represents ground level and the z-axis is vertically upwards.
The unit of length is the metre.

A thin pipe, U, runs directly below and parallel to the y-axis at a depth of 2 m below ground level.

i) Write down a vector equation for the line representing the pipe U.

A second straight thin pipe, V, links the point A(-4, 6, 0) on the ground with a point B on the pipe U. The vector equation of V is

$$\underline{r} = \begin{pmatrix} -4 \\ 6 \\ 0 \end{pmatrix} + t \begin{pmatrix} 2 \\ 1 \\ -1 \end{pmatrix}$$

where t is a parameter.

ii) Find the co-ordinates of B.

It is intended to lay a third straight pipeline, W, to connect the point C(12, 22, 0) on ground level with the first pipe at D(0, 0, -2).

iii) Prove that the pipe W can be laid without meeting the line occupied by the pipe V.
iv) Calculate the acute angle between the pipes U and W, giving your answer to the nearest degree.

5 Rational Functions

The purpose of this chapter is to enable you to

- simplify rational functions by factorising and cancelling or by polynomial division
- add and subtract rational functions
- express simple rational functions as a sum of partial fractions

Simplifying Rational Functions

The following are all examples of rational functions:

$$f(x) = \frac{3x - 2}{x^2 + 2}; \quad g(x) = \frac{x^4 - 2}{5x^6 - 20}; \quad h(x) = \frac{x^4 - 2}{x^2 + 2x - 8}$$

A **rational function** is a function whose rule is of the form $f(x) = \dfrac{p(x)}{q(x)}$ where p and q are polynomial functions. $p(x)$ is the numerator and $q(x)$ is the denominator of $f(x)$.

A rational function is said to be 'top heavy' if the degree of the numerator is greater than or equal to the degree of the denominator.

There are two ways in which a rational function may be simplified:

- the numerator and denominator can be factorised and any common factors may be cancelled out;
- 'top heavy' rational functions may be simplified by polynomial division.

> Recall that the degree of a polynomial $p(x)$ is the highest power of x in the polynomial.
>
> For example the degree of
> $$p(x) = 3x^5 - 7x^8 + 5x - 2$$
> is 8.

Simplification by Cancelling Out Common Factors

Recall that arithmetic fractions are simplified by cancelling out common factors of the numerator and denominator. For example

$$\frac{18}{42} = \frac{3 \times 6}{7 \times 6} = \frac{3 \times \cancel{6}^{1}}{7 \times \cancel{6}_{1}} = \frac{3}{7}.$$

In the same way, some rational functions can be simplified by fully factorising both the numerator and the denominator and then cancelling out any common factors.

EXAMPLE 1

Simplify $\dfrac{6x^2 - 12x}{x^3 - 4x}$.

EXAMPLE 1 (continued)

$$\frac{6x^2 - 12x}{x^3 - 4x} = \frac{6x(x-2)}{x(x^2 - 4)}$$

The denominator is not yet fully factorised.

$$= \frac{6x(x-2)}{x(x-2)(x+2)}$$

Both the numerator and the denominator have common factors of x and $(x-2)$.

$$= \frac{6^1 \cancel{x}(\cancel{x-2})}{\cancel{x}_1(\cancel{x-2}_1)(x+2)}$$

$$= \frac{6}{x+2}.$$

EXAMPLE 2

Simplify $\dfrac{2x - 12}{x^2 - 12x + 36}$.

$$\frac{2x - 12}{x^2 - 12x + 36} = \frac{2(x-6)}{(x-6)^2}$$

$$= \frac{2(\cancel{x-6})^1}{(x-6)(\cancel{x-6}_1)}$$

$$= \frac{2}{x-6}.$$

Simplification of Top Heavy Rational Functions by Polynomial Division

In module C2 simple cases of polynomial division were considered where the numerator was at most a cubic polynomial and the denominator was a linear polynomial. Two different methods were suggested and they are both shown in the next example.

EXAMPLE 3

Use polynomial division to simplify $\dfrac{3x^2 + 7x - 3}{x - 4}$.

Method 1

$$\frac{3x^2 + 7x - 3}{x - 4} = \frac{3x(x-4) + 12x + 7x - 3}{x - 4}$$

Since $3x^2 = 3x(x - 4) + 12x.$

$$= \frac{3x(x-4) + 19x - 3}{x - 4}$$

$$= 3x + \frac{19x - 3}{x - 4}$$

$$= 3x + \frac{19(x-4) + 76 - 3}{x - 4}$$

Since $19x = 19(x - 4) + 76.$

$$= 3x + 19 + \frac{73}{x - 4}.$$

The **quotient** when $3x^2 + 7x - 3$ is divided by $x - 4$ is $3x + 19$.

The **remainder** when $3x^2 + 7x - 3$ is divided by $x - 4$ is 73.

EXAMPLE 3 (continued)

Method 2

The second method depended on the observation that when a polynomial f is divided by a polynomial of degree 1 the quotient is a polynomial of degree one less than the degree of f and the remainder is simply a number.

The polynomial $3x^2 + 7x - 3$ is of degree 2 so when it is divided by $x - 4$, the quotient will be a polynomial of degree 1 and the remainder will be a number.
We therefore look for numbers P, Q and R so that

$$\frac{3x^2 + 7x - 3}{x - 4} \equiv Px + Q + \frac{R}{x - 4}.$$

Multiplying each term by the denominator, $(x - 4)$, gives

$$3x^2 + 7x - 3 \equiv Px(x - 4) + Q(x - 4) + R$$
$$\Rightarrow \quad 3x^2 + 7x - 3 = Px^2 + (Q - 4P)x + R - 4Q.$$

Looking at coefficients of x^2 $\quad\Rightarrow\quad$ $3 = P$.
Looking at coefficients of x $\quad\Rightarrow\quad$ $7 = Q - 4P$ $\quad\Rightarrow\quad$ $Q = 19$.
Looking at the constant coefficient $\quad\Rightarrow\quad$ $-3 = R - 4Q$ $\quad\Rightarrow\quad$ $R = 73$.

These three results give

$$\frac{3x^2 + 7x - 3}{x - 4} \equiv 3x + 19 + \frac{73}{x - 4}.$$

These two methods can be extended to deal with more complicated polynomial divisions. At this stage you will need to be able to handle numerators which are at most degree 4 and denominators which are at most degree 2.

EXAMPLE 4

Use polynomial division to simplify $\dfrac{2x^4 + 3x}{x^2 + 5}$.

Method 1

$$\frac{2x^4 + 3x}{x^2 + 5} = \frac{2x^2(x^2 + 5) - 10x^2 + 3x}{x^2 + 5}$$
Since $2x^4 = 2x^2(x^2 + 5) - 10x^2$.

$$= 2x^2 + \frac{-10x^2 + 3x}{x^2 + 5}$$

$$= 2x^2 + \frac{-10(x^2 + 5) + 50 + 3x}{x^2 + 5}$$
Since $-10x^2 = -10(x^2 + 5) + 50$.

$$= 2x^2 - 10 + \frac{3x + 50}{x^2 + 5}.$$

The **quotient** when $2x^4 + 3x$ is divided by $x^2 + 5$ is $2x^2 - 10$.

The **remainder** when $2x^4 + 3x$ is divided by $x^2 + 5$ is $3x + 50$.

EXAMPLE 4 (continued)

Method 2

The second method depends on the fact that when a polynomial f is divided by a polynomial of degree 2 then the quotient will be a polynomial of degree two less than the degree of f and the remainder will be a linear polynomial.

The polynomial $2x^4 + 3x$ is of degree 4 so when it is divided by $x^2 + 5$ the quotient will be a polynomial of degree 2 and the remainder will be a linear polynomial.

There will be values P, Q, R, S and T such that

$$\frac{2x^4 + 3x}{x^2 + 5} \equiv Px^2 + Qx + R + \frac{Sx + T}{x^2 + 5}.$$

Multiplying each term by the denominator, $(x^2 + 5)$, gives

$$2x^4 + 3x \equiv Px^2(x^2 + 5) + Qx(x^2 + 5) + R(x^2 + 5) + Sx + T$$
$$\implies \quad 2x^4 + 3x \equiv Px^4 + Qx^3 + (5P + R)x^2 + (5Q + S)x + 5R + T.$$

Looking at coefficients of x^4	$\implies$	$2 = P$.		
Looking at coefficients of x^3	$\implies$	$0 = Q$.		
Looking at coefficients of x^2	$\implies$	$0 = 5P + R$	$\implies$	$R = -10$.
Looking at coefficients of x	$\implies$	$3 = 5Q + S$	$\implies$	$S = 3$.
Looking at the constant coefficient	$\implies$	$0 = 5R + T$	$\implies$	$T = 50$.

These results give $\dfrac{2x^4 + 3x}{x^2 + 5} \equiv 2x^2 - 10 + \dfrac{3x + 50}{x^2 + 5}.$

If you are asked to simplify a rational function then it is best to first look to simplify by cancelling out common factors and then, if necessary, use polynomial division.

EXAMPLE 5

Simplify $\dfrac{x^3 - 9x}{x^2 + 2x - 3}$.

Start by looking for any common factors:

$$\frac{x^3 - 9x}{x^2 + 2x - 3} = \frac{x(x^2 - 9)}{(x + 3)(x - 1)}$$

$$= \frac{x(x + 3)(x - 3)}{(x + 3)(x - 1)}$$

It is important to factorise **fully** both the numerator and the denominator.

$$= \frac{x(x + 3)(x - 3)}{(x + 3)(x - 1)}$$

$$= \frac{x(x - 3)}{x - 1}.$$

EXAMPLE 5 (continued)

This fraction is still top heavy so polynomial division will further simplify the fraction:

$$\frac{x^3 - 9x}{x^2 + 2x - 3} = \frac{x(x-3)}{x-1}$$

$$= \frac{x^2 - 3x}{x-1}$$

$$= \frac{x(x-1) + x - 3x}{x-1}$$

$$= x + \frac{-2x}{x-1}$$

$$= x + \frac{-2(x-1) - 2}{x-1}$$

$$= x - 2 - \frac{2}{x-1}.$$

The Arithmetic of Rational Functions

Rational terms can be added, subtracted, multiplied and divided in exactly the same way as numerical fractions.

EXAMPLE 6

Write $\dfrac{5}{x} + \dfrac{3}{x+2}$ as a single fraction.

$$\frac{5}{x} + \frac{3}{x+2} = \frac{5(x+2)}{x(x+2)} + \frac{3x}{x(x+2)}$$

A common denominator for the two fractions is $x(x+2)$.

$$= \frac{5(x+2) + 3x}{x(x+2)}$$

$$= \frac{8x + 10}{x(x+2)}.$$

EXAMPLE 7

Write $\dfrac{4}{x-2} - \dfrac{3}{x+1} + \dfrac{1}{(x+1)^2}$ as a single fraction.

A common denominator for the three fractions is $(x-2)(x+1)^2$.

$$\frac{4}{x-2} - \frac{3}{x+1} + \frac{1}{(x+1)^2} = \frac{4(x+1)^2}{(x-2)(x+1)^2} - \frac{3(x-2)(x+1)}{(x-2)(x+1)^2} + \frac{1(x-2)}{(x-2)(x+1)^2}$$

$$= \frac{4(x^2 + 2x + 1) - 3(x^2 - x - 2) + (x-2)}{(x-2)(x+1)^2}$$

$$= \frac{x^2 + 12x + 8}{(x-2)(x+1)^2}.$$

EXERCISE 1

1 Simplify the following rational expressions:

a) $\dfrac{9x-27}{x^2-9}$ **b)** $\dfrac{x^2+8x+15}{x^3+5x^2}$ **c)** $\dfrac{5x^3-45x}{x^4-5x^2-36}$

2 Use polynomial division to simplify:

a) $\dfrac{x^2+3x+5}{x+2}$ **b)** $\dfrac{2x^3+14x+12}{x^2+6x}$ **c)** $\dfrac{x^3}{x-2}$

d) $\dfrac{3x^4+x^3-5}{x^2+4}$ **e)** $\dfrac{x^4+1}{x^2+2x+4}$

3 Simplify:

a) $\dfrac{x^3-16x}{2x^2+6x-8}$ **b)** $\dfrac{x^4-4x^3}{x^3-16x}$

4 Write the following expressions as single fractions:

a) $\dfrac{4}{x+1}+\dfrac{5}{x-2}$ **b)** $\dfrac{7}{x+3}-\dfrac{5}{x+2}$

c) $\dfrac{1}{x+3}+\dfrac{3}{x-2}+\dfrac{1}{x+1}$ **d)** $\dfrac{2}{x+5}-\dfrac{1}{x+2}+\dfrac{2}{x+1}$

e) $\dfrac{3}{x+2}+\dfrac{4}{x-1}+\dfrac{1}{(x-1)^2}$ **f)** $\dfrac{4}{x+5}-\dfrac{3}{2x-2}+\dfrac{1}{(x-5)^2}$

Partial Fractions

In chapter 2 we encountered partial fractions for the first time.

We saw how expressions of the form $\dfrac{Ax+B}{(x+a)(x+b)}$ could be rewritten as $\dfrac{P}{x+a}+\dfrac{Q}{x+b}$ and that this was of great use in both integrating rational expressions and finding the expansions of rational functions.

EXAMPLE 8

Write $\dfrac{x+4}{(x+1)(x+2)}$ as a sum of partial fractions and hence evaluate

$$\int_0^1 \dfrac{x+4}{(x+1)(x+2)}\,dx.$$

Suppose $\dfrac{x+4}{(x+1)(x+2)}=\dfrac{P}{x+1}+\dfrac{Q}{x+2}$.

Multiplying through by the common denominator, $(x+1)(x+2)$

$\Rightarrow \quad x+4=P(x+2)+Q(x+1).$

Putting $x=-2 \quad \Rightarrow \quad 2=0P-Q \quad \Rightarrow \quad Q=-2.$
Putting $x=-1 \quad \Rightarrow \quad 3=P+0Q \quad \Rightarrow \quad P=3.$

EXAMPLE 8 (continued)

Then

$$\Rightarrow \quad \frac{x+4}{(x+1)(x+2)} = \frac{3}{x+1} - \frac{2}{x+2}$$

so

$$\int_0^1 \frac{x+4}{(x+1)(x+2)}\, dx = \int_0^1 \frac{3}{x+1} - \frac{2}{x+2}\, dx$$
$$= \Big[3\ln(x+1) - 2\ln(x+2)\Big]_0^1$$
$$= (3\ln 2 - 2\ln 3) - (3\ln 1 - 2\ln 2)$$
$$= 5\ln 2 - 2\ln 3.$$

Fractions with Three Distinct Linear Factors

The expression $\dfrac{4}{1+x} + \dfrac{2}{1-2x} - \dfrac{3}{2-x}$ can be written as a single fraction by using $(1+x)(1-2x)(2-x)$ as a common denominator.

$$\frac{4}{1+x} + \frac{2}{1-2x} - \frac{3}{2-x} \equiv \frac{4(1-2x)(2-x)}{(1+x)(1-2x)(2-x)} + \frac{2(1+x)(2-x)}{(1+x)(1-2x)(2-x)} - \frac{3(1+x)(1-2x)}{(1+x)(1-2x)(2-x)}$$

$$\equiv \frac{4(2-5x+2x^2) + 2(2+x-x^2) - 3(1-x-2x^2)}{(1+x)(1-2x)(2-x)}$$

$$\equiv \frac{9-15x+12x^2}{(1+x)(1-2x)(2-x)}.$$

The experience gained from this example suggests that a fraction of the form $\dfrac{Ax^2+Bx+C}{(ax+b)(cx+d)(ex+f)}$ is the result of adding fractions of the form $\dfrac{P}{ax+b}$, $\dfrac{Q}{cx+d}$ and $\dfrac{R}{ex+f}$.

EXAMPLE 9

SOLUTION

Write $\dfrac{4x^2-3x+8}{(x+1)(x+2)(4-x)}$ as a sum of partial fractions.

We want to find values P, Q and R so that

$$\frac{4x^2-3x+8}{(x+1)(x+2)(4-x)} \equiv \frac{P}{x+1} + \frac{Q}{x+2} + \frac{R}{4-x}. \tag{1}$$

Since

$$\frac{P}{x+1} + \frac{Q}{x+2} + \frac{R}{4-x} \equiv \frac{P(x+2)(4-x) + Q(x+1)(4-x) + R(x+1)(x+2)}{(x+1)(x+2)(4-x)}$$

values P, Q and R must be found so that

$$\frac{4x^2-3x+8}{(x+1)(x+2)(4-x)} \equiv \frac{P(x+2)(4-x) + Q(x+1)(4-x) + R(x+1)(x+2)}{(x+1)(x+2)(4-x)}.$$

EXAMPLE 9 (continued)

Since the two identical fractions have the same denominator, their numerators must be identical:

$$4x^2 - 3x + 8 \equiv P(x + 2)(4 - x) + Q(x + 1)(4 - x) + R(x + 1)(x + 2). \qquad [2]$$

This identity must hold for **all values of x**. In particular the identity must be valid when $x = -2$, when $x = 4$ and when $x = -1$.

> In practice, we will usually move directly from identity [1] through to identity [2] by multiplying equation [1] by the common denominator of the fraction to be expanded.

Putting $x = -2$ into [2] gives

$$30 = P \times 0 + Q \times -6 + R \times 0 \quad \Rightarrow \quad Q = -5.$$

Putting $x = 4$ into [2] gives

$$60 = P \times 0 + Q \times 0 + R \times 30 \quad \Rightarrow \quad R = 2.$$

> These values are chosen since they enable the values of P, Q and R to be determined immediately from equation [2].

Putting $x = -1$ into [2] gives

$$15 = P \times 5 + Q \times 0 + R \times 0 \quad \Rightarrow \quad P = 3.$$

We have $\dfrac{4x^2 - 3x + 8}{(x + 1)(x + 2)(4 - x)} \equiv \dfrac{3}{x + 1} - \dfrac{5}{x + 2} + \dfrac{2}{4 - x}.$

EXAMPLE 10

The function f is defined by $f(x) = \dfrac{6 + 11x - 26x^2}{(2 + x)(1 + 2x)(1 - 2x)}$.

a) Express $f(x)$ as a sum of partial fractions.
b) Find the first four terms in the expansion, in ascending powers of x, of $f(x)$. State the values of x for which the expansion is valid.

a) Values P, Q and R are needed so that

$$\frac{6 + 11x - 26x^2}{(2 + x)(1 + 2x)(1 - 2x)} = \frac{P}{2 + x} + \frac{Q}{1 + 2x} + \frac{R}{1 - 2x}.$$

Multiplying through by $(2 + x)(1 + 2x)(1 - 2x)$ gives

$$6 + 11x - 26x^2 \equiv P(1 + 2x)(1 - 2x) + Q(2 + x)(1 - 2x) + R(2 + x)(1 + 2x).$$

Putting $x = -\dfrac{1}{2}$ gives $-6 = 3Q$ $\Rightarrow$ $Q = -2.$

Putting $x = \dfrac{1}{2}$ gives $5 = 5R$ $\Rightarrow$ $R = 1.$

Putting $x = -2$ gives $-120 = -15P$ $\Rightarrow$ $P = 8.$

This gives $f(x) = \dfrac{6 + 11x - 26x^2}{(2 + x)(1 + 2x)(1 - 2x)} = \dfrac{8}{2 + x} - \dfrac{2}{1 + 2x} + \dfrac{1}{1 - 2x}.$

EXAMPLE 10 (continued)

b) Recall from chapter 3 that if $-1 < z < 1$ then we can write

$$(1+z)^n = 1 + nz + \frac{n(n-1)}{2!}z^2 + \frac{n(n-1)(n-2)}{3!}z^3 + \cdots + \frac{n(n-1)\dots(n-r+1)}{r!}z^r + \dots .$$

Putting $n = -1$ gives

$$(1+z)^{-1} = 1 + (-1)z + \frac{-1((-1)-1)}{2!}z^2 + \frac{-1((-1)-1)((-1)-2)}{3!}z^3 + \dots$$

$$= 1 - z + z^2 - z^3 + \dots \qquad \text{provided} \quad -1 < z < 1.$$

The expansions for $\dfrac{1}{2+x}, \dfrac{1}{1+2x}, \dfrac{1}{1-2x}$ can be deduced from this:

$$\frac{1}{2+x} = (2+x)^{-1}$$

$$= \left(2\left(1 + \frac{1}{2}x\right)\right)^{-1}$$

$$= 2^{-1}\left(1 + \frac{1}{2}x\right)^{-1}$$

$$= \frac{1}{2}\left(1 - \left(\frac{1}{2}x\right) + \left(\frac{1}{2}x\right)^2 - \left(\frac{1}{2}x\right)^3 + \dots\right) \qquad \text{provided} \quad -1 < \frac{1}{2}x < 1$$

$$= \frac{1}{2} - \frac{1}{4}x + \frac{1}{8}x^2 - \frac{1}{16}x^3 + \dots \qquad \text{provided} \quad -2 < x < 2$$

$$\frac{1}{1+2x} = (1+2x)^{-1}$$

$$= (1 - (2x) + (2x)^2 - (2x)^3 + \dots) \qquad \text{provided} \quad -1 < 2x < 1$$

$$= 1 - 2x + 4x^2 - 8x^3 + \dots \qquad \text{provided} \quad -\frac{1}{2} < x < \frac{1}{2}$$

$$\frac{1}{1-2x} = (1 + (-2x))^{-1}$$

$$= (1 - (-2x) + (-2x)^2 - (-2x)^3 + \dots) \qquad \text{provided} \quad -1 < -2x < 1$$

$$= 1 + 2x + 4x^2 + 8x^3 + \dots \qquad \text{provided} \quad -\frac{1}{2} < x < \frac{1}{2}.$$

Finally,

$$f(x) = \frac{6 + 11x - 26x^2}{(2+x)(1+2x)(1-2x)} = \frac{8}{2+x} - \frac{2}{1+2x} + \frac{1}{1-2x}$$

$$= 8\left(\frac{1}{2} - \frac{1}{4}x + \frac{1}{8}x^2 - \frac{1}{16}x^3 + \dots\right) - 2(1 - 2x + 4x^2 - 8x^3 + \dots) + (1 + 2x + 4x^2 + 8x^3 + \dots)$$

$$= 4 - 2x + x^2 - \frac{1}{2}x^3 - 2 + 4x - 8x^2 + 16x^3 + 1 + 2x + 4x^2 + 8x^3 + \dots$$

$$= 3 + 4x - 3x^2 + \frac{47}{2}x^3 + \dots .$$

For this series to be valid, x must satisfy both $-2 < x < 2$ and $-\frac{1}{2} < x < \frac{1}{2}$.

This is the case when $-\frac{1}{2} < x < \frac{1}{2}$.

EXERCISE 2

1 Express the following as sums of partial fractions:

a) $\dfrac{7x+14}{(x-3)(x+4)}$ **b)** $\dfrac{x+23}{(x-2)(x+3)}$ **c)** $\dfrac{3}{x^2+2x-8}$

2 Evaluate $\displaystyle\int_0^2 \dfrac{3-x}{(x+3)(x+1)}\,dx.$

3 a) Express $\dfrac{3x+18}{(1+x)(2-x)}$ as a sum of partial fractions.

b) Hence prove that the first four terms in the expansion in ascending powers of x of

$\dfrac{3x+18}{(1+x)(2-x)}$ are $9-3x+6x^2-\dfrac{9}{2}x^3$.

State the values of x for which this expansion is valid.

4 Express the following as sums of partial fractions:

a) $\dfrac{x^2-12x+11}{(x-3)(x-2)(x+1)}$ **b)** $\dfrac{35x^2+5x}{(x+1)(2x-1)(3x+1)}$

5 Express $\dfrac{2x^2+11x+8}{(x+1)(x+2)(2x+1)}$ as a sum of partial fractions.

Hence prove that $\displaystyle\int_0^1 \dfrac{2x^2+11x+8}{(x+1)(x+2)(2x+1)}\,dx = 3\ln 2.$

Fractions with a Repeated Linear Factor

The expression $\dfrac{3}{1+2x}+\dfrac{1}{(1+2x)^2}-\dfrac{3}{2-x}$ can be written as a single fraction by using $(1+2x)^2(2-x)$ as a common denominator:

$$\dfrac{3}{1+2x}+\dfrac{1}{(1+2x)^2}-\dfrac{3}{2-x} = \dfrac{3(1+2x)(2-x)}{(1+2x)^2(2-x)}+\dfrac{1(2-x)}{(1+2x)^2(2-x)}-\dfrac{3(1+2x)^2}{(1+2x)^2(2-x)}$$

$$\equiv \dfrac{3(2+3x-2x^2)+(2-x)-3(1+4x+4x^2)}{(1+2x)^2(2-x)}$$

$$\equiv \dfrac{5-4x-18x^2}{(1+2x)^2(2-x)}.$$

This suggests that a fraction of the form $\dfrac{Ax^2+Bx+C}{(ax+b)^2(cx+d)}$ is the result of adding fractions of the form $\dfrac{P}{ax+b}$, $\dfrac{Q}{(ax+b)^2}$ and $\dfrac{R}{ex+f}$.

EXAMPLE 11

Write $\dfrac{4x^2 - 8x - 3}{(1+x)(2-x)^2}$ as a sum of partial fractions.

We want values P, Q and R so that

$$\frac{4x^2 - 8x - 3}{(1+x)(2-x)^2} \equiv \frac{P}{1+x} + \frac{Q}{2-x} + \frac{R}{(2-x)^2}.$$

Multiplying through by the common denominator:

$$4x^2 - 8x - 3 \equiv P(2-x)^2 + Q(1+x)(2-x) + R(1+x).$$

Putting $x = 2$ $\implies$ $-3 = 3R$ $\implies$ $R = -1$.
Putting $x = -1$ $\implies$ $9 = 9P$ $\implies$ $P = 1$.
Putting $x = 0$ $\implies$ $-3 = 4P + 2Q + R$ $\implies$ $Q = -3$.

This gives

$$\frac{4x^2 - 8x - 3}{(1+x)(2-x)^2} \equiv \frac{1}{1+x} - \frac{3}{2-x} - \frac{1}{(2-x)^2}.$$

We have now seen how to express three types of fraction as a sum of partial fractions:

$$\frac{Ax + B}{(ax+b)(cx+d)} \quad \text{can be written as} \quad \frac{P}{ax+b} + \frac{Q}{cx+d}$$

$$\frac{Ax^2 + Bx + C}{(ax+b)(cx+d)(ex+f)} \quad \text{can be written as} \quad \frac{P}{ax+b} + \frac{Q}{cx+d} + \frac{R}{ex+f}$$

$$\frac{Ax^2 + Bx + C}{(ax+b)^2(cx+d)} \quad \text{can be written as} \quad \frac{P}{ax+b} + \frac{Q}{(ax+b)^2} + \frac{R}{cx+d}$$

> You need to learn these results.

The values of P, Q and R can be found by multiplying through by the common denominator and then substituting suitable values of x

EXAMPLE 12

a) Write $\dfrac{12x + 32}{(x+3)^2(x+1)}$ as a sum of partial fractions.

b) Hence find the exact value of

$$\int_0^3 \frac{12x + 32}{(x+3)^2(x+1)}\, dx.$$

> Don't be put off by the lack of an x^2 term in the numerator: you can always say
> $$\frac{12x + 32}{(x+3)^2(x+1)} = \frac{0x^2 + 12x + 32}{(x+3)^2(x+1)}.$$

a) We want values P, Q and R so that

$$\frac{12x + 32}{(x+3)^2(x+1)} \equiv \frac{P}{x+3} + \frac{Q}{(x+3)^2} + \frac{R}{x+1}.$$

Multiplying through by the common denominator

$$\implies \quad 12x + 32 \equiv P(x+3)(x+1) + Q(x+1) + R(x+3)^2.$$

EXAMPLE 12 (continued)

We want this to be valid for **all values of** x, so

putting $x = -3$ $\Rightarrow$ $-4 = -2Q$ $\Rightarrow$ $Q = 2$
putting $x = -1$ $\Rightarrow$ $20 = 4R$ $\Rightarrow$ $R = 5$
putting $x = 0$ $\Rightarrow$ $32 = 3P + Q + 9R$ $\Rightarrow$ $32 = 3P + 2 + 45$
$\Rightarrow$ $P = -5.$

Then $\dfrac{12x + 32}{(x+3)^2(x+1)} \equiv \dfrac{-5}{x+3} + \dfrac{2}{(x+3)^2} + \dfrac{5}{x+1}$

b) Using the result of (a) we can write

$$\int_0^3 \frac{12x+32}{(x+3)^2(x+1)} \, dx = \int_0^3 \left(-5 \times \frac{1}{x+3} + 2 \times \frac{1}{(x+3)^2} + 5 \times \frac{1}{x+1} \right) dx.$$

The integral of $\dfrac{1}{x+3}$ is $\ln|x+3|$.

The integral of $\dfrac{1}{x+1}$ is $\ln|x+1|$.

Using the substitution $u = x + 3$,

$$\int \frac{1}{(x+3)^2} \, dx = \int \frac{1}{u^2} \, du$$

$$= \int u^{-2} \, du$$

$$= \frac{u^{-1}}{-1}$$

$$= -\frac{1}{u}$$

$$= -\frac{1}{x+3}.$$

If $u = x + 3$ then

$$\frac{1}{(x+3)^2} = \frac{1}{u^2}$$

and

$$\frac{du}{dx} = 1 \Rightarrow \text{'}dx = du\text{'}.$$

Putting these three results together gives

$$\int_0^3 \frac{12x+32}{(x+3)^2(x+1)} \, dx = \left[-5\ln|x+3| - 2 \times \frac{1}{x+3} + 5\ln|x+1| \right]_0^3$$

$$= \left(-5\ln 6 - \frac{1}{3} + 5\ln 4 \right) - \left(-5\ln 3 - \frac{2}{3} + 5\ln 1 \right)$$

$$= 5(\ln 4 + \ln 3 - \ln 6) + \frac{2}{3} - \frac{1}{3}$$

$$= 5\ln\left(\frac{4 \times 3}{6}\right) + \frac{1}{3}$$

$$= 5\ln 2 + \frac{1}{3}.$$

EXERCISE 3

1 Write the following expressions as a sum of partial fractions:

a) $\dfrac{5x^2 + 24x + 32}{(x-2)(x+3)^2}$

b) $\dfrac{9x}{(2x+1)^2(1-x)}$

c) $\dfrac{4x^2 - x - 2}{x^2(x+1)}$

d) $\dfrac{7x^2 + 16x + 1}{(x-5)(x+3)^2}$

e) $\dfrac{8x^2 - 14x + 8}{x^2(2-x)}$

2 a) Write down $\displaystyle\int \dfrac{1}{1+x}\,dx$ and $\displaystyle\int \dfrac{1}{2+x}\,dx$.

b) By making the substitution $u = 2 + x$ evaluate $\displaystyle\int \dfrac{1}{(2+x)^2}\,dx$.

c) Express $f(x) = \dfrac{3x^2 + 11x + 9}{(1+x)(2+x)^2}$ as a sum of partial fractions.

d) Evaluate $\displaystyle\int_0^1 f(x)\,dx$.

3 Find $\displaystyle\int \dfrac{19 - 6x + 2x^2}{(x+1)(x-2)^2}\,dx$.

4 a) Write down the first four terms of the binomial expansions for

i) $(1 + x)^{-1}$ ii) $(1 + x)^{-2}$ iii) $(1 - x)^{-1}$

b) Express $h(x) = \dfrac{4 - 5x - 3x^2}{(1-x)(1+x)^2}$ as a sum of partial fractions.

c) Show that the first four terms of the expansion of $h(x)$ in ascending powers of x are
$4 - 9x + 10x^2 - 15x^3 + \dots$.

5 Express $g(x) = \dfrac{5 - 6x}{(1+2x)(1-2x)^2}$ as a sum of partial fractions.
Hence obtain the first five terms in the expansion of $g(x)$ and state the values of x for which the series is valid.

6 Express $f(x) = \dfrac{20 - 2x - 9x^2}{(1-x)(2-x)(1+2x)}$ as a sum of partial fractions.
Hence find

a) $\displaystyle\int f(x)\,dx$,

b) the expansion, in ascending powers of x, up to the x^3 term, stating the range of values for which the expansion is valid.

Having studied this chapter you should know

- how to simplify a rational function by factorising the numerator and denominator and cancelling any common factors
- how to simplify a top heavy rational function by polynomial division
- that $\dfrac{Ax+B}{(ax+b)(cx+d)}$ can be written as $\dfrac{P}{ax+b}+\dfrac{Q}{cx+d}$
- that $\dfrac{Ax^2+Bx+C}{(ax+b)(cx+d)(ex+f)}$ can be written as $\dfrac{P}{ax+b}+\dfrac{Q}{cx+d}+\dfrac{R}{ex+f}$
- that $\dfrac{Ax^2+Bx+C}{(ax+b)^2(cx+d)}$ can be written as $\dfrac{P}{ax+b}+\dfrac{Q}{(ax+b)^2}+\dfrac{R}{cx+d}$
- how to use partial fractions to integrate some rational functions
- how to use partial fractions to produce expansions of some rational functions

REVISION EXERCISE

1 Simplify $\dfrac{x^2+7x-30}{x^3-9x}$.

2 Express $\dfrac{3x}{(x-2)(x+1)}$ as a sum of partial fractions.

Hence find the exact value of $\displaystyle\int_3^4 \dfrac{3x}{(x-2)(x+1)}\,dx$.

3 a) Find values P, Q, R and S so that $\dfrac{3x^3+10x^2-8x-10}{(x-1)(x+4)}=Px+Q+\dfrac{R}{x-1}+\dfrac{S}{x+4}$.

b) Hence find the value of $\dfrac{d^2y}{dx^2}$ when $x-2$ if $y=\dfrac{3x^3+10x^2-8x-10}{(x-1)(x+4)}$.

4 Express $\dfrac{3x-7}{(x+1)(x+2)}$ as a sum of partial fractions.

Hence evaluate $\displaystyle\int_3^4 \dfrac{3x-7}{(x+1)(x+2)}\,dx$.

5 i) Given that

$$y=\dfrac{4x^2-13x+12}{(1-x)^2(2-x)}$$

express y as a sum of three partial fractions.

ii) The expression for y is to be expanded in ascending powers of x, where $-1<x<1$. Find the terms of the expansion up to and including the term in x^3.

6 **a)** Simplify $\dfrac{4x^2 - 8x}{x^3 + 2x^2 - 8x}$.

   b) Express $\dfrac{4x + 1}{x^2 - 4} + \dfrac{2}{x + 3}$ as a single fraction.

   c) Use polynomial division to simplify $\dfrac{x^4 - 3x^2}{x^2 + 1}$.

7 **i)** Express $\dfrac{1}{x(x - 4)(x + 4)}$ in partial fractions.

   ii) Hence find the exact value of $\displaystyle\int_1^3 \dfrac{1}{x(x - 4)(x + 4)}\,dx$, giving your answer in the form

   $\lambda \ln\left(\dfrac{a}{b}\right)$, where a and b are integers.

(OCR Jan 2002 P3)

8 Find integers A, B, C and D such that

$$\frac{4x^3 + 8x^2 + x - 3}{x(1 + 2x)} = Ax + B + \frac{C}{x} + \frac{D}{1 + 2x}$$

and hence find $\displaystyle\int \frac{4x^3 + 8x^2 + x - 3}{x(1 + 2x)}\,dx$.

9 Express $f(x) = \dfrac{9x^2 - 7x + 2}{(1 + x)(1 - x)(1 - 2x)}$ as a sum of three partial fractions.

The expression for $f(x)$ is to be expanded in ascending powers of x. Find the terms of the expansion up to and including the term in x^3 and state the values of x for which the expansion is valid.

10 **i)** Find the expansion in ascending powers of x, up to the x^6 term, of $\dfrac{1}{9 - x^2}$.

   ii) Write $g(x) = \dfrac{1 + x}{3 + x} - \dfrac{1 - x}{3 - x}$ as a single fraction.

   iii) Obtain the expansion in ascending powers of x, up to the x^7 term, of $g(x)$.

11 Find the quotient and remainder when $3x^3 + 5x^2 - 4x - 5$ is divided by $x^2 + 2x - 2$.

12 Determine the values of λ and μ if

$$\frac{5}{x + 1} + \frac{\lambda}{(x - 2)^2} - \frac{3}{x - 2} \equiv \frac{2x^2 - \mu x + 34}{(x + 1)(x - 2)^2}.$$

6 Techniques of Integration 2

The purpose of this chapter is to enable you to

- use integration by parts to integrate products of functions
- use integration by substitution to integrate a variety of functions
- build the experience necessary to determine which method of integration should be used

Integration by Parts

The product rule for differentiation (chapter 3 of C3) states that $\dfrac{d}{dx}(uv) = \dfrac{du}{dx}v + u\dfrac{dv}{dx}$.

This can be rewritten as the integration statement:

$$\int \left(\frac{du}{dx}v + u\frac{dv}{dx} \right) dx = uv$$

$$\Rightarrow \quad \int \frac{du}{dx}v\,dx + \int u\frac{dv}{dx}\,dx = uv$$

$$\Rightarrow \quad \int u\frac{dv}{dx}\,dx = uv - \int \frac{du}{dx}v\,dx.$$

This formula is one way of expressing the **integration by parts** formula which will be seen to be useful for integrating some expressions which are products of two functions.

As an initial example, consider the problem of finding $\displaystyle\int x\,e^{2x}\,dx$.

If we write

$$u = x \quad \text{and} \quad \frac{dv}{dx} = e^{2x}$$

then

> We will see later that it is permissible to omit the integration constant at this stage.

$$\frac{du}{dx} = 1 \quad \text{and} \quad v = \int e^{2x}\,dx = \frac{1}{2}e^{2x}.$$

When these are substituted into the integration by parts formula

$$\int u\frac{dv}{dx}\,dx = uv - \int \frac{du}{dx}v\,dx$$

we obtain

$$\int x\,e^{2x}\,dx = x\left(\frac{1}{2}e^{2x} \right) - \int 1 \times \left(\frac{1}{2}e^{2x} \right) dx$$

$$= \frac{1}{2}x\,e^{2x} - \int \left(\frac{1}{2}e^{2x} \right) dx$$

> Observe that the $\displaystyle\int \frac{du}{dx}v\,dx$ integral is much simpler to evaluate than the original $\displaystyle\int u\frac{dv}{dx}\,dx$ integral.

$$= \frac{1}{2}x\,e^{2x} - \frac{1}{4}e^{2x}.$$

To complete the integration we must remember that an indefinite integral requires an unknown constant in the final solution. We write the final result as

$$\int x\, e^{2x}\, dx = \frac{1}{2}x\, e^{2x} - \frac{1}{4}e^{2x} + c.$$

Now consider the problem of finding $\int x^3 \ln x\, dx$.

If we write

$$u = x^3 \quad \text{and} \quad \frac{dv}{dx} = \ln x$$

then

$$\frac{du}{dx} = 3x^2$$

but it is not possible to find an expression for v, since $v = \int \ln x\, dx$ and we do not know how to integrate $\ln x$.

However, since $\int x^3 \ln x\, dx = \int (\ln x)x^3\, dx$, we can write

$$u = \ln x \quad \text{and} \quad \frac{dv}{dx} = x^3$$

$$\Rightarrow \quad \frac{du}{dx} = \frac{1}{x} \quad \text{and} \quad v = \frac{1}{4}x^4.$$

When these are substituted into the integration by parts formula

$$\int u\frac{dv}{dx}\, dx = uv - \int \frac{du}{dx}v\, dx$$

we obtain

$$\int (\ln x)x^3\, dx = (\ln x)\left(\frac{1}{4}x^4\right) - \int \frac{1}{x} \times \left(\frac{1}{4}x^4\right) dx$$

$$= \frac{1}{4}x^4 \ln x - \int \left(\frac{1}{4}x^3\right) dx$$

$$= \frac{1}{4}x^4 \ln x - \frac{1}{16}x^4 + c.$$

> Observe that the $\int \frac{du}{dx}v\, dx$ integral is much simpler to evaluate than the original $\int u\frac{dv}{dx}\, dx$ integral.

To use integration by parts to evaluate $\int f(x)g(x)\,dx$

1 Let $u = f(x)$ and $\dfrac{dv}{dx} = g(x)$

2 Obtain $\dfrac{du}{dx}$ and $v = \int g(x)\,dx.$

$\left[\text{If } v \text{ cannot be found try putting } u = g(x) \text{ and } \dfrac{dv}{dx} = f(x)\right]$

3 Substitute the values in the integration by parts formula:

$$\int u\frac{dv}{dx}\,dx = uv - \int \frac{du}{dx}v\,dx.$$

If possible, evaluate the $\int \dfrac{du}{dx}v\,dx$ integral. Don't forget the integration constant

The integration by parts formula is useful when integrating a product of two different types of function: for example a polynomial and a trigonometric, exponential or logarithmic function.

EXAMPLE 1

SOLUTION

Find $\int x \cos 3x\,dx.$

If

$$u = x \quad \text{and} \quad \frac{dv}{dx} = \cos 3x$$

then

$$\frac{du}{dx} = 1 \quad \text{and} \quad v = \int \cos 3x\,dx = \frac{1}{3}\sin 3x.$$

The integration by parts formula

$$\int u\frac{dv}{dx}\,dx = uv - \int \frac{du}{dx}v\,dx$$

gives

$$\int x \cos 3x\,dx = x\left(\frac{1}{3}\sin 3x\right) - \int 1 \times \left(\frac{1}{3}\sin 3x\right)dx$$

$$= \frac{1}{3}x \sin 3x - \int \frac{1}{3}\sin 3x\,dx$$

$$= \frac{1}{3}x \sin 3x - \left(-\frac{1}{9}\cos 3x\right) + c$$

$$= \frac{1}{3}x \sin 3x + \frac{1}{9}\cos 3x + c.$$

You may have noticed that the integration constants have been omitted until the final line of the working of an integration by parts example. You may wonder why this is possible? In particular, why is it permissible to drop the integration constant when writing down v?

To answer this, consider example 1 again but this time including the integration constant **every time** an integration is done.

If $\quad u = x \quad$ and $\quad \dfrac{dv}{dx} = \cos 3x$

then

$$\dfrac{du}{dx} = 1 \quad \text{and} \quad v = \int \cos 3x \, dx = \dfrac{1}{3} \sin 3x + k.$$

> k is the integration constant for this first integration.

The integration by parts formula

$$\int u \dfrac{dv}{dx} dx = uv - \int \dfrac{du}{dx} v \, dx$$

gives

$$\int x \cos 3x \, dx = x \left(\dfrac{1}{3} \sin 3x + k \right) - \int 1 \times \left(\dfrac{1}{3} \sin 3x + k \right) dx$$

$$= \dfrac{1}{3} x \sin 3x + kx - \int \left(\dfrac{1}{3} \sin 3x + k \right) dx$$

$$= \dfrac{1}{3} x \sin 3x + kx - \left(-\dfrac{1}{9} \cos 3x + kx \right) + c$$

> c is the integration constant for the second integration.

$$= \dfrac{1}{3} x \sin 3x + kx + \dfrac{1}{9} \cos 3x - kx + c$$

$$= \dfrac{1}{3} x \sin 3x + \dfrac{1}{9} \cos 3x + c.$$

No terms involving the first integration constant, k, appear in the final answer. Therefore the integration constant for the first integration can be safely omitted.

Sometimes it may be necessary to use integration by parts more than once to find an integral:

EXAMPLE 2

Find $\int x^2 e^{-2x} \, dx$.

If

$$u = x^2 \quad \text{and} \quad \dfrac{dv}{dx} = e^{-2x}$$

then

$$\dfrac{du}{dx} = 2x \quad \text{and} \quad v = \int e^{-2x} \, dx = -\dfrac{1}{2} e^{-2x}.$$

The integration by parts formula

$$\int u \dfrac{dv}{dx} dx = uv - \int \dfrac{du}{dx} v \, dx$$

gives

$$\int x^2 e^{-2x} \, dx = x^2 \times \left(-\dfrac{1}{2} e^{-2x} \right) - \int (2x) \times \left(\dfrac{1}{2} e^{-2x} \right) dx$$

$$= -\dfrac{1}{2} x^2 e^{-2x} + \int x e^{-2x} \, dx. \hspace{3cm} [1]$$

EXAMPLE 2 (continued)

To progress we need to find $\int x\,e^{-2x}\,dx$. A second application of the integration by parts formula is required:

If

$$u = x \quad \text{and} \quad \frac{dv}{dx} = e^{-2x}$$

then

$$\frac{du}{dx} = 1 \quad \text{and} \quad v = \int e^{-2x}\,dx = -\frac{1}{2}e^{-2x}.$$

The integration by parts formula

$$\int u\frac{dv}{dx}\,dx = uv - \int \frac{du}{dx}\,v\,dx$$

gives

$$\int x\,e^{-2x}\,dx = x \times \left(-\frac{1}{2}e^{-2x}\right) - \int 1 \times \left(-\frac{1}{2}e^{-2x}\right)dx$$

$$= -\frac{1}{2}x\,e^{-2x} + \int \frac{1}{2}e^{-2x}\,dx$$

$$= -\frac{1}{2}x\,e^{-2x} - \frac{1}{4}e^{-2x}. \qquad\qquad [2]$$

Combining results [1] and [2] gives

$$\int x^2\,e^{-2x}\,dx = -\frac{1}{2}x^2\,e^{-2x} + \int x\,e^{-2x}\,dx$$

$$= -\frac{1}{2}x^2\,e^{-2x} + \left(-\frac{1}{2}x\,e^{-2x} - \frac{1}{4}e^{-2x}\right) + c$$

$$= -\frac{1}{2}x^2\,e^{-2x} - \frac{1}{2}x\,e^{-2x} - \frac{1}{4}e^{-2x} + c.$$

Evaluating Definite Integrals using Integration by Parts

Definite integrals can be evaluated by first calculating the indefinite integral and then inserting the limits at the end of the question.

Alternatively, since uv is the integral of $\frac{du}{dx}v + u\frac{dv}{dx}$ we can write

$$\int_a^b \left(\frac{du}{dx}v + u\frac{dv}{dx}\right)dx = [uv]_a^b$$

and rearranging this gives the integration by parts result for definite integrals:

$$\Rightarrow \quad \int_a^b u\frac{dv}{dx}\,dx = [uv]_a^b - \int_a^b \frac{du}{dx}\,v\,dx.$$

These two approaches are illustrated in example 3.

EXAMPLE 3

Find the exact value of $\int_0^4 x\, e^{\frac{1}{2}x}\, dx$.

If

$$u = x \quad \text{and} \quad \frac{dv}{dx} = e^{\frac{1}{2}x}$$

then

$$\frac{du}{dx} = 1 \quad \text{and} \quad v = \int e^{\frac{1}{2}x}\, dx = 2e^{\frac{1}{2}x}.$$

The integration by parts formula

$$\int u \frac{dv}{dx}\, dx = uv - \int \frac{du}{dx} v\, dx$$

gives

$$\int x\, e^{\frac{1}{2}x}\, dx = x\left(2e^{\frac{1}{2}x}\right) - \int 1 \times \left(2e^{\frac{1}{2}x}\right) dx$$

$$= 2x\, e^{\frac{1}{2}x} - \int 2e^{\frac{1}{2}x}\, dx$$

$$= 2x\, e^{\frac{1}{2}x} - 4e^{\frac{1}{2}x} + c.$$

Inserting the limits, and remembering that the integration constant can be dropped in a definite integral, gives

$$\int_0^4 x\, e^{\frac{1}{2}x}\, dx = \left[2x\, e^{\frac{1}{2}x} - 4e^{\frac{1}{2}x}\right]_0^4$$

$$= \left(8e^2 - 4e^2\right) - (0 - 4)$$

$$= 4e^2 + 4.$$

If

$$u = x \quad \text{and} \quad \frac{dv}{dx} = e^{\frac{1}{2}x}$$

then

$$\frac{du}{dx} = 1 \quad \text{and} \quad v = \int e^{\frac{1}{2}x}\, dx = 2e^{\frac{1}{2}x}.$$

The integration by parts formula for definite integrals

$$\int_a^b u \frac{dv}{dx}\, dx = [uv]_a^b - \int_a^b \frac{du}{dx} v\, dx$$

gives

$$\int_0^4 x\, e^{\frac{1}{2}x}\, dx = \left[x\left(2e^{\frac{1}{2}x}\right)\right]_0^4 - \int_0^4 1 \times \left(2e^{\frac{1}{2}x}\right) dx$$

$$= \left[2x\, e^{\frac{1}{2}x}\right]_0^4 - \int_0^4 2e^{\frac{1}{2}x}\, dx$$

$$= \left(8e^2 - 0\right) - \left[4e^{\frac{1}{2}x}\right]_0^4$$

$$= 8e^2 - \left(4e^2 - 4\right)$$

$$= 4e^2 + 4.$$

EXAMPLE 4

Find $\displaystyle\int_1^{e^2} \sqrt{x}\,\ln x\,dx$.

We first calculate $\displaystyle\int \sqrt{x}\,\ln x\,dx$.

If

$$u = \ln x \quad \text{and} \quad \frac{dv}{dx} = \sqrt{x} = x^{\frac{1}{2}}$$

then

> Applying the integration by parts formula with $u = \sqrt{x}$ and $\dfrac{dv}{dx} = \ln x$ is not possible since the integral of $\ln x$ is not yet known.
>
> However, this can easily be remedied by switching the two functions over.

$$\frac{du}{dx} = \frac{1}{x} \quad \text{and} \quad v = \int x^{\frac{1}{2}}\,dx = \frac{2}{3}x^{\frac{3}{2}}.$$

The integration by parts formula

$$\int u\frac{dv}{dx}\,dx = uv - \int \frac{du}{dx}\,v\,dx$$

gives

$$\int \sqrt{x}\,\ln x\,dx = \int (\ln x)\sqrt{x}\,dx$$

$$= (\ln x)\left(\frac{2}{3}x^{\frac{3}{2}}\right) - \int \frac{1}{x}\times\left(\frac{2}{3}x^{\frac{3}{2}}\right)dx$$

$$= \frac{2}{3}x^{\frac{3}{2}}\ln x - \int \frac{2}{3}x^{\frac{1}{2}}\,dx$$

$$= \frac{2}{3}x^{\frac{3}{2}}\ln x - \frac{2}{3}\times\frac{2}{3}x^{\frac{3}{2}} + c$$

$$= \frac{2}{3}x^{\frac{3}{2}}\ln x - \frac{4}{9}x^{\frac{3}{2}} + c.$$

Inserting the limits, and remembering that the integration constant can be dropped in a definite integral, gives

$$\int_1^{e^2} \sqrt{x}\,\ln x\,dx = \left[\frac{2}{3}x^{\frac{3}{2}}\ln x - \frac{4}{9}x^{\frac{3}{2}}\right]_1^{e^2}$$

$$= \left(\frac{2}{3}(e^2)^{\frac{3}{2}}\ln(e^2) - \frac{4}{9}(e^2)^{\frac{3}{2}}\right) - \left(\frac{2}{3}(1)^{\frac{3}{2}}\ln(1) - \frac{4}{9}(1)^{\frac{3}{2}}\right)$$

$$= \left(\frac{4}{3}e^3 - \frac{4}{9}e^3\right) - \left(0 - \frac{4}{9}\right)$$

> $\ln(e^2) = 2\ln e = 2$;
> $\ln 1 = 0$.

$$= \frac{8}{9}e^3 + \frac{4}{9}.$$

EXERCISE 1

Use integration by parts to find:

1 $\displaystyle\int x\cos x\,dx$

2 $\displaystyle\int xe^{5x}\,dx$

3 $\displaystyle\int x\cos 3x\,dx$

4 $\displaystyle\int x^4\ln x\,dx$

5 $\displaystyle\int x\sin x\,dx$

6 $\displaystyle\int (2x+3)\,e^x\,dx$

7 $\displaystyle\int x^2\cos x\,dx$

8 $\displaystyle\int x^2\sin 2x\,dx$

9 $\displaystyle\int x\sec^2 x\,dx$

10 $\displaystyle\int_1^2 x^2\ln x\,dx$

11 $\displaystyle\int_0^1 x^2 e^{3x}\,dx$

12 $\displaystyle\int_1^2 xe^{-2x}\,dx$

13 By writing $\ln x = (\ln x)\times 1$ find $\displaystyle\int \ln x\,dx$.

14 Sketch the graph of the function $y = y^2 e^{-x}$ taking care to show the position of the stationary points.
Prove that the area of the region bounded by the x-axis, the line $x = 2$ and the curve is $2 - 10e^{-2}$.

15 The curve $y = \sqrt{x}e^x$ between $x = 0$ and $x = 2$ is rotated completely around the x-axis.
Prove that the resulting solid has volume $\dfrac{\pi}{4}\left(3e^4 + 1\right)$.

16 Find the exact value of the volume of the solid generated when the curve $y = x\sqrt{\sin x}$ between $x = 0$ and $x = \pi$ is rotated completely around the x-axis.

17 Let $J = \displaystyle\int e^{2x}\cos 3x\,dx$.

a) Using integration by parts **twice**, prove that

$$J = \frac{1}{3}e^{2x}\sin 3x + \frac{2}{9}e^{2x}\cos 3x - \frac{4}{9}J$$

and hence find J.

b) Use a similar method to determine the value of $\displaystyle\int e^{3x}\sin 5x\,dx$.

Integration by Substitution

We have already seen how very natural substitutions can greatly simplify the evaluation of lots of integrals.

EXAMPLE 5

Find $\int \dfrac{x}{(x-3)^2}\, dx$.

$$\int \frac{x}{(x-3)^2}\, dx = \int \frac{u+3}{u^2}\, du$$

$$= \int \frac{1}{u} + \frac{3}{u^2}\, du$$

$$= \ln|u| - \frac{3}{u} + c$$

$$= \ln|x-3| - \frac{3}{x-3} + c.$$

If the substitution $u = x - 3$ is used then

$$(x-3)^2 = u^2,$$
$$x = u + 3$$

and

$$\frac{du}{dx} = 1 \Rightarrow 'dx = du'.$$

Choosing a substitution that will help in the evaluation of an integral is essentially a matter of experience. A good starting principle is to try to simplify the most complicated part of the function to be integrated.

Recall that the easiest way of handling definite integrals which require substitution is to change the limits of the integration as you proceed.

EXAMPLE 6

Evaluate $\displaystyle\int_0^2 x\sqrt{2x^2+1}\, dx$.

$$\int_0^2 x\sqrt{2x^2+1}\, dx = \int_1^9 \frac{1}{4} u^{\frac{1}{2}}\, du$$

$$= \left[\frac{1}{4} \times \frac{2}{3} u^{\frac{3}{2}}\right]_1^9$$

$$= \left(\frac{1}{6} \times 9^{\frac{3}{2}}\right) - \left(\frac{1}{6} \times 1^{\frac{3}{2}}\right)$$

$$= \frac{9}{2} - \frac{1}{6}$$

$$= \frac{13}{3}.$$

Making the substitution

$$u = 2x^2 + 1$$

will simplify the square root part of the function to be integrated. If this substitution is used then

$$\sqrt{2x^2+1} = \sqrt{u} = u^{\frac{1}{2}}$$

and

$$\frac{du}{dx} = 4x \Rightarrow 'dx = \frac{1}{4x}\, du'$$

so

$$x\sqrt{2x^2+1}\, dx = xu^{\frac{1}{2}} \frac{1}{4x}\, du = \frac{1}{4} u^{\frac{1}{2}}\, du.$$

Considering the limits

$$x = 0 \implies u = 2 \times 0^2 + 1 = 1$$
$$x = 2 \implies u = 2 \times 2^2 + 1 = 9.$$

EXAMPLE 7

Find $\int \sin^3 x \cos x \, dx$.

SOLUTION

$$\int \sin^3 x \cos x \, dx = \int u^3 \, du$$

$$= \frac{1}{4} u^4 + c$$

$$= \frac{1}{4} \sin^4 x + c.$$

Making the substitution

$$u = \sin x$$

will simplify the cube part of the function to be integrated. If this substitution is used then

$$\sin^3 x = u^3$$

and

$$\frac{du}{dx} = \cos x \implies 'dx = \frac{1}{\cos x} du'$$

so

$$\sin^3 x \cos x \, dx = u^3 \cos x \times \frac{1}{\cos x} du = u^3 \, du.$$

EXAMPLE 8

Evaluate $\int_0^{\frac{\pi}{4}} \sec^2 \theta \sqrt{4 + 5 \tan \theta} \, d\theta$.

SOLUTION

$$\int_0^{\frac{\pi}{4}} \sec^2 \theta \sqrt{4 + 5 \tan \theta} \, d\theta = \int_4^9 \frac{1}{5} u^{\frac{1}{2}} \, du$$

$$= \left[\frac{2}{15} u^{\frac{3}{2}} \right]_4^9$$

$$= \left(\frac{2}{15} \times 9^{\frac{3}{2}} \right) - \left(\frac{2}{15} \times 4^{\frac{3}{2}} \right)$$

$$= \frac{2}{15} \times 27 - \frac{2}{15} \times 8$$

$$= \frac{38}{15}.$$

Making the substitution

$$u = 4 + 5 \tan \theta$$

will simplify the square root part of the function to be integrated. If this substitution is used then

$$\sqrt{4 + 5 \tan \theta} = \sqrt{u} = u^{\frac{1}{2}}$$

and

$$\frac{du}{d\theta} = 5 \sec^2 \theta \implies 'd\theta = \frac{1}{5 \sec^2 \theta} du'$$

so

$$\sec^2 \theta \sqrt{4 + 5 \tan \theta} \, d\theta = \sec^2 \theta \times u^{\frac{1}{2}} \times \frac{1}{5 \sec^2 \theta} du$$

$$= \frac{1}{5} u^{\frac{1}{2}} \, du.$$

Considering the limits:

$$\theta = 0 \implies u = 4 + 5 \tan 0 = 4$$

$$\theta = \frac{\pi}{4} \implies u = 4 + 5 \tan\left(\frac{\pi}{4}\right) = 9.$$

EXERCISE 2 — Work through this exercise carefully. It is important to build up experience of which substitutions are likely to prove helpful in evaluating integrals.

1 **a)** Use the substitution $u = 4x - 3$ to find $\int (4x - 3)^5 \, dx$.

b) Use suitable substitutions to find

i) $\int_0^2 \sqrt{4x + 1} \, dx$ **ii)** $\int \dfrac{6}{(2x - 1)^4} \, dx$

2 **a)** Use the substitution $u = x^2 + 4$ to find $\int 4x(x^2 + 4)^5 \, dx$.

b) Use suitable substitutions to find

i) $\int \dfrac{8x}{(x^2 + 25)} \, dx$ **ii)** $\int_0^4 t\sqrt{t^2 + 9} \, dt$ **iii)** $\int x^3(x^2 + 4)^5 \, dx$

c) Explain why the substitution $u = x^2 + 1$ does not help in finding the integrals

$$\int \sqrt{x^2 + 1} \, dx \quad \text{and} \quad \int x^2\sqrt{x^2 + 1} \, dx.$$

3 **a)** Use the substitution $u = x^3 + 2$ to evaluate $\int \dfrac{x^2}{(x^3 + 2)^3} \, dx$.

b) Use suitable substitutions to find

i) $\int \dfrac{6x^2}{\sqrt{x^3 + 4}} \, dx$ **ii)** $\int x^4(x^5 + 4)^{10} \, dx$ **iii)** $\int_0^1 t^3\sqrt{7t^4 + 9} \, dt$ **iv)** $\int x^9(x^5 + 4)^{10} \, dx$

4 **a)** Use the substitution $u = 1 + e^{2x}$ to find $\int \dfrac{e^{2x}}{\sqrt{1 + e^{2x}}} \, dx$.

b) Use suitable substitutions to find

i) $\int e^x(1 + 5e^x)^3 \, dx$ **ii)** $\int \dfrac{e^{-2x}}{(4 + e^{-2x})^2} \, dx$

5 **a)** Use the substitution $u = \sin \theta$ to evaluate $\int_0^{\frac{\pi}{2}} \sin^5 \theta \cos \theta \, d\theta$.

b) Use suitable substitutions to find

i) $\int \cos^4 x \sin x \, dx$ **ii)** $\int \tan^7 2x \sec^2 2x \, dx$

6 **a)** Prove that $\sin^5 x \cos^3 x \equiv (\sin^5 x - \sin^7 x)\cos x$.

b) Use the substitution $u = \sin x$ to evaluate $\int_0^{\frac{\pi}{2}} \sin^5 x \cos^3 x \, dx$.

c) Evaluate $\int_0^{\frac{\pi}{2}} \sin^3 x \cos^8 x \, dx$.

7 Evaluate the following integrals:

a) $\int_0^{\frac{\pi}{2}} \dfrac{\cos \theta}{(4 + \sin \theta)^3} \, d\theta$ **b)** $\int_0^3 (2x - 1)^3 \, dx$ **c)** $\int_0^1 3x(x^2 + 1)^3 \, dx$

d) $\int_0^{\frac{\pi}{2}} \sin^3 x \cos x \, dx$ **e)** $\int_1^e \dfrac{(\ln x)^3}{x} \, dx$ **f)** $\int_0^3 \sqrt{x + 1} \, dx$

Further Substitutions

The following examples illustrate some substitutions that you may find useful in future work.

EXAMPLE 9

Find $\int \sin^6 \theta \cos^3 \theta \, d\theta$ by making the substitution $u = \sin \theta$.

SOLUTION

$$\int \sin^6 \theta \cos^3 \theta \, d\theta = \int u^6 (1 - u^2) \, du$$

$$= \int (u^6 - u^8) \, du$$

$$= \frac{1}{7} u^7 - \frac{1}{9} u^9 + c$$

$$= \frac{1}{7} \sin^7 \theta - \frac{1}{9} \sin^9 \theta + c.$$

If
$$u = \sin \theta$$
then
$$\sin^6 \theta = u^6$$
and
$$\frac{du}{d\theta} = \cos \theta \Rightarrow \text{'d}\theta = \frac{1}{\cos \theta} \, du\text{'}$$
so
$$\sin^6 \theta \cos^3 \theta \, d\theta = u^6 \cos^3 \theta \times \frac{1}{\cos \theta} \, du$$
$$= u^6 \cos^2 \theta \, du$$
$$= u^6 (1 - \sin^2 \theta) \, du$$
$$= u^6 (1 - u^2) \, du$$
since $\cos^2 \theta = 1 - \sin^2 \theta = 1 - u^2$.

- The substitution $u = \sin \theta$ will prove useful in finding $\int \sin^p \theta \cos^q \theta \, d\theta$ **provided q is an odd positive integer**.

- The substitution $u = \cos \theta$ will prove useful in finding $\int \sin^p \theta \cos^q \theta \, d\theta$ **provided p is an odd positive integer**.

EXAMPLE 10

Use the substitution $u = 9 - x^2$ to find

a) $\int x^3 \sqrt{9 - x^2} \, dx$ **b)** $\int \frac{x}{\sqrt{9 - x^2}} \, dx$

SOLUTION

a) $\int x^3 \sqrt{9 - x^2} \, dx = \int -\frac{1}{2} (9 - u) u^{\frac{1}{2}} \, du$

$$= \frac{1}{2} \int \left(u^{\frac{3}{2}} - 9 u^{\frac{1}{2}} \right) du$$

$$= \frac{1}{2} \left(\frac{2}{5} u^{\frac{5}{2}} - 9 \times \frac{2}{3} u^{\frac{3}{2}} \right) + c$$

$$= \frac{1}{5} u^{\frac{5}{2}} - 3 u^{\frac{3}{2}} + c$$

$$= \frac{1}{5} \left(9 - x^2 \right)^{\frac{5}{2}} - 3 \left(9 - x^2 \right)^{\frac{3}{2}} + c.$$

If
$$u = 9 - x^2$$
then
$$\sqrt{9 - x^2} = \sqrt{u} = u^{\frac{1}{2}}$$
and
$$\frac{du}{dx} = -2x \Rightarrow \text{'d}x = -\frac{1}{2x} \, du\text{'}$$
so
$$x^3 \sqrt{9 - x^2} \, dx = x^3 u^{\frac{1}{2}} \times \left(-\frac{1}{2x} \, du \right)$$
$$= -\frac{1}{2} x^2 u^{\frac{1}{2}} \, du$$
$$= -\frac{1}{2} (9 - u) u^{\frac{1}{2}} \, du$$
since $u = 9 - x^2 \Rightarrow x^2 = 9 - u$.

EXAMPLE 10 (continued)

b) $\int \dfrac{x}{\sqrt{9-x^2}}\,dx = \int -\dfrac{1}{2}u^{-\frac{1}{2}}\,du$

$$= -u^{\frac{1}{2}} + c$$

$$= -\sqrt{u} + c$$

$$= -\sqrt{9-x^2} + c.$$

If

$$u = 9 - x^2$$

then

$$\sqrt{9-x^2} = \sqrt{u} = u^{\frac{1}{2}}$$

and

$$\frac{du}{dx} = -2x \implies \text{'}dx = -\frac{1}{2x}\,du\text{'}$$

so

$$\frac{x}{\sqrt{9-x^2}}\,dx = \frac{x}{\sqrt{u}} \times \left(-\frac{1}{2x}\,du\right)$$

$$= -\frac{1}{2\sqrt{u}}\,du$$

$$= -\frac{1}{2}u^{-\frac{1}{2}}\,du.$$

- The substitution $u = a^2 - x^2$ will prove useful in finding $\int x^p\sqrt{a^2-x^2}\,dx$ or $\int \dfrac{x^p}{\sqrt{a^2-x^2}}\,dx$ provided **p is an odd positive integer**.

EXAMPLE 11

Use the substitution $x = 4\sin\theta$ to evaluate $\displaystyle\int_0^2 \dfrac{x^2}{\sqrt{16-x^2}}\,dx$.

SOLUTION

$$\int_0^2 \frac{x^2}{\sqrt{16-x^2}}\,dx = \int_0^{\frac{\pi}{6}} 16\sin^2\theta\,d\theta.$$

In chapter 1 we saw that the identity

$$\sin^2 A = \frac{1}{2} - \frac{1}{2}\cos 2A$$

was useful in integrating $\sin^2\theta$.

This means that

$$\int_0^2 \frac{x^2}{\sqrt{16-x^2}}\,dx = \int_0^{\frac{\pi}{6}} 16\sin^2\theta\,d\theta$$

$$= \int_0^{\frac{\pi}{6}} 16\left(\frac{1}{2} - \frac{1}{2}\cos 2\theta\right)d\theta$$

$$= \int_0^{\frac{\pi}{6}} (8 - 8\cos 2\theta)\,d\theta$$

$$= \left[8\theta - 4\sin 2\theta\right]_0^{\pi/6}$$

$$= \left(\frac{8\pi}{6} - 4\sin\left(\frac{\pi}{3}\right)\right) - (0 - 0)$$

$$= \frac{4\pi}{3} - 2\sqrt{3}.$$

If

$$x = 4\sin\theta$$

then

$$\sqrt{16-x^2} = \sqrt{16 - 16\sin^2\theta}$$

$$= 4\sqrt{1 - \sin^2\theta}$$

$$= 4\sqrt{\cos^2\theta}$$

$$= 4\cos\theta$$

and

$$\frac{dx}{d\theta} = 4\cos\theta \implies \text{'}dx = 4\cos\theta\,d\theta\text{'}$$

so

$$\frac{x^2}{\sqrt{16-x^2}}\,dx = \frac{16\sin^2\theta}{4\cos\theta} \times (4\cos\theta\,d\theta)$$

$$= 16\sin^2\theta\,d\theta.$$

Considering the limits:

$$x = 0 \implies 4\sin\theta = 0 \implies \theta = 0.$$

$$x = 2 \implies 4\sin\theta = 2$$

$$\implies \sin\theta = \frac{1}{2} \implies \theta = \frac{\pi}{6}.$$

- The substitution $x = a \sin \theta$ will prove useful in evaluating $\int x^p \sqrt{a^2 - x^2}\, dx$ or $\int \dfrac{x^p}{\sqrt{a^2 - x^2}}\, dx$ if p **is an even positive integer or zero**.

EXERCISE 3

Evaluate the following integrals:

1 $\displaystyle\int_2^3 x(2x-3)^5\, dx$

2 $\displaystyle\int_0^{\frac{\pi}{2}} \cos^4 x \sin^3 x\, dx$

3 $\displaystyle\int_0^3 \dfrac{x^2}{\sqrt{9-x^2}}\, dx$

4 $\displaystyle\int_2^3 \dfrac{x^2}{(x-1)^3}\, dx$

5 $\displaystyle\int_0^1 \dfrac{x^3}{\sqrt{1+x^2}}\, dx$

6 $\displaystyle\int_0^5 \sqrt{25-x^2}\, dx$

7 $\displaystyle\int_0^{\frac{\pi}{2}} \sin^3 x\, dx$

8 $\displaystyle\int_0^{\frac{\pi}{2}} \cos^3 x \sin^2 x\, dx$

9 **a)** Prove that $\sin^2 A \cos^2 A \equiv \dfrac{1}{8} - \dfrac{1}{8}\cos 4A$.

  b) Hence evaluate

  i) $\displaystyle\int_0^{\frac{\pi}{2}} \sin^2 2x \cos^2 2x\, dx$ **ii)** $\displaystyle\int_0^4 x^2\sqrt{16-x^2}\, dx$

10 Evaluate

  i) $\displaystyle\int_0^1 \dfrac{x}{\sqrt{4-x^2}}\, dx$ **ii)** $\displaystyle\int_0^1 \dfrac{x^2}{\sqrt{4-x^2}}\, dx$

11 Find $\displaystyle\int \dfrac{1}{\sqrt{1-x^2}}\, dx$.
  Deduce the derivative of $\sin^{-1} x$.

12 **a)** Use the substitution $x = \tan \theta$ to find $\displaystyle\int \dfrac{1}{1+x^2}\, dx$.

  b) Deduce the derivative of $\tan^{-1} x$.

  c) By writing $\tan^{-1} x$ as $(\tan^{-1} x) \times 1$, use integration by parts to determine $\displaystyle\int \tan^{-1} x\, dx$.

Integration: Which Method?

The method of differentiating a function is usually evident from the way in which the function is built up from simple functions.

On the other hand, deciding which method of integration a function requires is much more difficult. The table on the next page attempts to give a flow chart to help you decide which method of integration to use. In fact there is no substitute for experience – be patient and keep trying different methods until you find one that works!

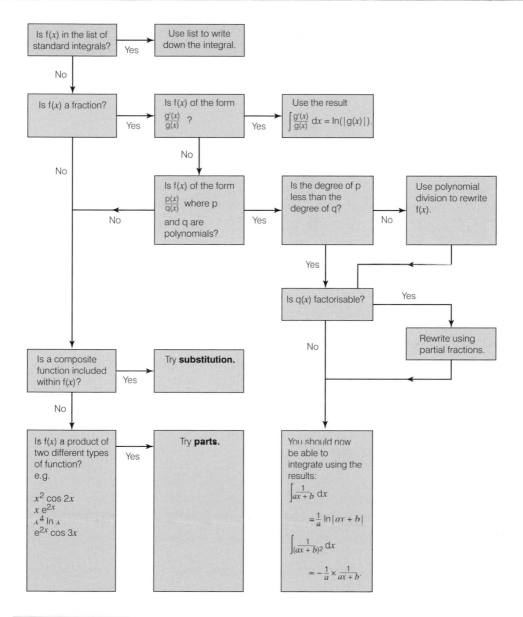

EXERCISE 4

1 $\displaystyle\int x(1+x^2)\,dx$

2 $\displaystyle\int xe^{4x}\,dx$

3 $\displaystyle\int_0^{\frac{\pi}{2}} \sin^2\theta\cos^3\theta\,d\theta$

4 $\displaystyle\int (e^{2x}-1)^2\,dx$

5 $\displaystyle\int \frac{\cos x}{\sqrt{2+\sin x}}\,dx$

6 $\displaystyle\int_3^4 \frac{5}{x^2+x-6}\,dx$

7 $\displaystyle\int_0^{\frac{\pi}{6}} \tan 2\theta\,d\theta$

8 $\displaystyle\int_0^5 x\sqrt{4+x}\,dx$

9 $\displaystyle\int_1^2 \frac{(x^2-1)^2}{x^3}\,dx$

10 $\displaystyle\int_0^1 t^2\sqrt{1-t}\,dt$

11 $\displaystyle\int_0^{\frac{\pi}{2}} x^2 \sin x\,dx$

12 $\displaystyle\int_0^1 \frac{8}{\sqrt{3+4x}}\,dx$

13 $\displaystyle\int \frac{x(2x+1)}{x+1}\,dx$

14 $\displaystyle\int_1^2 x^4 \ln x\,dx$

15 $\displaystyle\int_0^3 x\sqrt{x^2+16}\,dx$

16 $\displaystyle\int_0^{\frac{\pi}{2}} \sin\theta\cos^2\theta\,d\theta$

17 $\displaystyle\int \frac{x^2}{x+2}\,dx$

18 $\displaystyle\int_1^e \frac{\ln x}{\sqrt{x}}\,dx$

19 $\displaystyle\int \frac{2x^3}{x^4+1}\,dx$

20 $\displaystyle\int \frac{11x-16}{(x-2)^2(x+1)}\,dx$

21 $\displaystyle\int_0^2 \frac{x^2}{\sqrt{4-x^2}}\,dx$

22 $\displaystyle\int_0^1 (1-x)\sin x\,dx$

23 $\displaystyle\int \ln(2x)\,dx$

Having studied this chapter you should

- know how to use the integration by parts formula $\displaystyle\int u\frac{dv}{dx}\,dx = uv - \int \frac{du}{dx}v\,dx$
- be able to use a wide variety of substitutions to evaluate integrals
- be able to decide on the appropriate method of integration for a function

---

REVISION EXERCISE

1 Prove that $\displaystyle\int_0^1 xe^{-\frac{1}{2}x}\,dx = 4 - 6e^{-1}$.

2 If V denotes the volume of the solid of revolution generated when the curve $y = \cos\left(\frac{1}{2}x\right)$ between $x = 0$ and $x = 2\pi$ is rotated by 2π radians about the x-axis, prove that $V = \pi^2$.

3 Using the substitution $u = 1 + 3\cos x$, or otherwise, find $\displaystyle\int \sin x(1 + 3\cos x)^4\,dx$.

4 Find $\displaystyle\int x\sec^2 x\,dx$.

(OCR Jun 2002 P3)

5 Using the substitution $u = 9 + x^2$, or otherwise, evaluate $\displaystyle\int_0^4 \frac{x}{\sqrt{9+x^2}}\,dx$.

6 Evaluate $\displaystyle\int_0^{\frac{1}{6}\pi} x\sin 3x\,dx$.

7 Evaluate $\displaystyle\int_2^4 \frac{x}{(x-1)^2}\,dx$.

8 Expand $\cos(2\theta + \theta)$.
Deduce that $\cos 3\theta \equiv 4\cos^3\theta - 3\cos\theta$.

By using the substitution $u = \cos\theta$, evaluate $\displaystyle\int_0^{\frac{1}{3}\pi} \cos 3\theta \sin\theta \, d\theta$.

9 Express $\dfrac{x-4}{x(x-2)}$ as a sum of partial fractions.

Hence prove that $\displaystyle\int_3^4 \dfrac{x-4}{x(x-2)}\,dx = \ln\left(\dfrac{8}{9}\right)$.

10 i) By using the substitution $x = a\sin\theta$, show that

$$\int_{\frac{1}{2}a}^{a} \sqrt{a^2 - x^2}\,dx = \frac{a^2}{2}\left(\frac{\pi}{3} - \frac{\sqrt{3}}{4}\right).$$

ii) The diagram shows the circle

$$x^2 + y^2 = a^2$$

and the line

$$x = \frac{1}{2}a.$$

Find the area of the shaded region, giving your answer in an exact form.

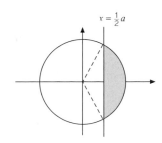

(OCR Jan 2002 P3)

11 The diagram shows a sketch of the curve $y = x\sqrt{\ln x}$.
The closed region bounded by this curve, the x-axis and the line $x = e$ is rotated completely about the x-axis. Prove that the resulting solid has volume $\dfrac{\pi}{9}(2e^3 + 1)$.

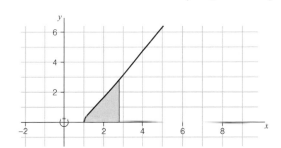

12 i) Use the derivative of $\cos x$ to prove that

$$\frac{d}{dx}(\sec x) = \sec x \tan x.$$

ii) Use the substitution $u = \sec x$ to find the exact value of

$$\int_0^{\frac{1}{3}\pi} \sec^3 x \tan^3 x \, dx.$$

(OCR Jun 2001 P3)

13 Express $\dfrac{9x^2}{(1+x)(2-x)^2}$ as a sum of partial fractions.

Prove that

$$\int_0^1 \frac{9x^2}{(1+x)(2-x)^2}\,dx = p + q\ln 2$$

where p and q are integers to be determined.

7 Parametric and Implicit Curves

The purpose of this chapter is to enable you to

- understand the use of a pair of parametric equations to define a curve, and use a given parametric representation of a curve in simple cases

- convert the equation of a curve between parametric and Cartesian forms

- understand the use of implicit equations to define a curve

- find and use the first derivative of a function which is defined parametrically or implicitly

The Parametric Representation of Curves

The curves that we have met so far have all been of the form $y = f(x)$. In some circumstances it is more convenient to define a curve by giving the co-ordinates of points (x, y) on the curve in terms of a third parameter, often denoted by t or θ.

EXAMPLE 1

Sketch the curve whose equation is given parametrically by
$$x = t^3, y = 6t^2.$$

SOLUTION

We can get an initial understanding of the curve by producing a table of values of x and y for various values of t:

t	−5	−4	−3	−2	−1	0	1	2	3	4	5
$x = t^3$	−125	−64	−27	−8	−1	0	1	8	27	64	125
$y = 3t^2$	150	96	54	24	6	0	6	24	54	96	150

These points can then be plotted to obtain

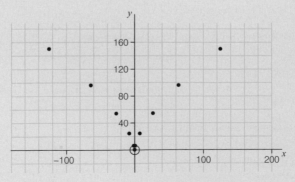

EXAMPLE 1 (continued)

The points can then be joined to obtain the curve with parametric equations

$$x = t^3, \ y = 6t^2.$$

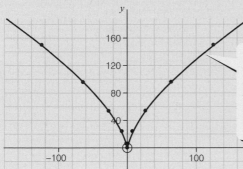

Parametric curves can easily be drawn on either a graphical calculator or using computer graph-drawing software. The main thing is to ensure that you allow the parameter to take a sufficiently large range of values to ensure that all the important features of the graph are evident.

EXAMPLE 2

A curve C is given parametrically by the equations

$$x = 5 \cos \theta + 2 \cos 5\theta$$
$$y = 5 \sin \theta + 2 \sin 5\theta.$$

Sketch the path of the point P.

This diagram was produced using a computer graph-drawing package.

Since $\cos \theta$ and $\sin \theta$ have a period of 2π whilst $\cos 5\theta$ and $\sin 5\theta$ have a period of $\dfrac{2\pi}{5}$, values of θ between 0 and 2π were considered.

SOLUTION

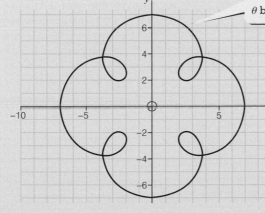

Finding the Cartesian Equation of a Curve whose Equation is Given Parametrically

With some parametric equations it is possible to convert the parametric equation of the curve into a Cartesian equation relating the variables x and y with no mention of the parameter. This usually requires some simple algebra and, possibly, some use of trigonometric results to eliminate the parameter from the equations.

Consider the curve of example 1, whose parametric equations were

$$x = t^3, \; y = 6t^2.$$

We can write

$$\frac{x}{y} = \frac{t^3}{6t^2} = \frac{t}{6}$$

$$\Rightarrow \quad t = \frac{6x}{y}.$$

> It is frequently very difficult or impossible to find a Cartesian equation for a curve given parametrically.
>
> For example, a Cartesian form for the curve whose parametric equations are
>
> $$x = 5 \cos \theta + 2 \cos 5\theta$$
> $$y = 5 \sin \theta + 2 \sin 5\theta$$
>
> cannot be easily found.

The equation $y = 6t^2$ can now be rewritten as

$$y = 6\left(\frac{6x}{y}\right)^2$$

$$\Rightarrow \quad y = \frac{216x^2}{y^2}$$

$$\Rightarrow \quad y^3 = 216x^2.$$

EXAMPLE 3

Sketch the curve with parametric equations
$$x = 5 \cos 2\theta, \; y = 3 \sin \theta.$$
Find a Cartesian equation for this curve.

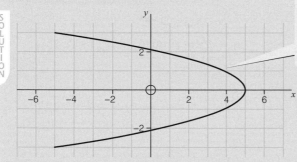

> This diagram was produced using a computer graph-drawing package. Since $\cos 2\theta$ has a period of π whilst $\sin \theta$ has a period of 2π, values of θ between 0 and 2π were considered.

To find a Cartesian equation for the curve, the parameter θ must be eliminated from the equations
$$x = 5 \cos 2\theta, \; y = 3 \sin \theta.$$
Recalling that $\cos 2\theta \equiv 1 - 2 \sin^2 \theta$, we can write

$$x = 5 \cos 2\theta$$
$$\Rightarrow \quad x = 5(1 - 2 \sin^2 \theta).$$

Since $y = 3 \sin \theta$, we know that $\sin \theta = \dfrac{y}{3}$

$$\Rightarrow \quad x = 5\left(1 - 2\left(\frac{y}{3}\right)^2\right)$$

$$\Rightarrow \quad x = 5\left(1 - \frac{2y^2}{9}\right)$$

$$\Rightarrow \quad x = \frac{5}{9}(9 - 2y^2).$$

> Notice that the parametric curve
> $$x = 5 \cos 2\theta, \; y = 3 \sin \theta$$
> only gives **part** of the curve $x = \dfrac{5}{9}(9 - 2y^2)$ since the x co-ordinate of the parametric curve can only take values between -5 and 5 and the y co-ordinate of the parametric curve can only take values between -3 and 3.

EXAMPLE 4

Sketch the curve with parametric equations

$$x = \ln t, \, y = \ln(3t^2)$$

and find a Cartesian equation for this curve.

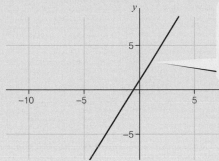

This diagram was produced using a computer graph-drawing package. Since $\ln t$ is only defined for **positive** values of t, the parameter t was allowed to vary between 0.01 and 100.

Using the properties of logarithms, we can write

$$
\begin{aligned}
y &= \ln(3t^2) \\
&= \ln 3 + \ln(t^2) \\
&= \ln 3 + 2 \ln t \qquad \text{Since } x = \ln t. \\
&= 2x + \ln 3
\end{aligned}
$$

so a Cartesian equation of the graph is $y = 2x + \ln 3$.

EXAMPLE 5

Sketch the curve with parametric equations

$$x = 3 + 5 \cos \theta, \, y = 1 + 3 \sin \theta$$

and find a Cartesian equation for this curve.

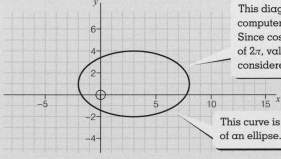

This diagram was produced using a computer graph-drawing package. Since $\cos \theta$ and $\sin \theta$ each have a period of 2π, values of θ between 0 and 2π were considered.

This curve is an example of an ellipse.

Rewriting the parametric equations as

$$\cos \theta = \frac{x - 3}{5} \quad \text{and} \quad \sin \theta = \frac{y - 1}{3}$$

and remembering that

$$\cos^2 \theta + \sin^2 \theta \equiv 1$$

EXAMPLE 5 (continued)

gives

$$\left(\frac{x-3}{5}\right)^2 + \left(\frac{y-1}{3}\right)^2 = 1$$

$$\Rightarrow \quad \frac{(x-3)^2}{25} + \frac{(y-1)^2}{9} = 1 \quad \text{or} \quad 9(x-3)^2 + 25(y-1)^2 = 225.$$

EXAMPLE 6

The diagram shows the curve with parametric equations

$$x = (2\pi - \theta)\cos\theta, \ y = \theta\sin\theta \qquad 0 \leqslant \theta \leqslant 2\pi.$$

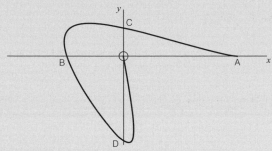

Find the co-ordinates of the points A, B, C and D.

At A and B, $y = 0$

$$\Rightarrow \quad \theta\sin\theta = 0$$
$$\Rightarrow \quad \theta = 0 \quad \text{or} \quad \sin\theta = 0$$
$$\Rightarrow \quad \theta = 0, \pi \quad \text{or} \quad 2\pi.$$

When $\theta = 0$, $x = (2\pi - 0)\cos 0 = 2\pi$.

Since A is the only point where $y = 0$ and $x > 0$, we deduce that A is the point $(2\pi, 0)$.

When $\theta = \pi$, $x = (2\pi - \pi)\cos\pi = -\pi$.

Since B is the only point where $y = 0$ and $x < 0$, we deduce that B is the point $(-\pi, 0)$.

When $\theta = 2\pi$, $x = (2\pi - 2\pi)\cos 2\pi = 0$, so $\theta = 2\pi$ gives the origin.

At C and D, $x = 0$

$$\Rightarrow \quad (2\pi - \theta)\cos\theta = 0$$
$$\Rightarrow \quad 2\pi - \theta = 0 \quad \text{or} \quad \cos\theta = 0$$
$$\Rightarrow \quad \theta = 2\pi \quad \text{or} \quad \theta = \frac{\pi}{2} \quad \text{or} \quad \frac{3\pi}{2}.$$

We already know that $\theta = 2\pi$ gives the origin.

When $\theta = \frac{\pi}{2}$, $y = \frac{\pi}{2}\sin\left(\frac{\pi}{2}\right) = \frac{\pi}{2}$.

Since C is the only point where $x = 0$ and $y > 0$, we deduce that C is the point $\left(0, \frac{\pi}{2}\right)$.

When $\theta = \frac{3\pi}{2}$, $y = \frac{3\pi}{2}\sin\left(\frac{\pi}{2}\right) = -\frac{3\pi}{2}$.

Since D is the only point where $x = 0$ and $y < 0$, we deduce that D is the point $\left(0, -\frac{3\pi}{2}\right)$.

EXERCISE 1

A graphical calculator will help greatly in producing the sketches required in this exercise.

1 Sketch the curve whose parametric equations are $x = t^2, y = 2t$ and find a Cartesian equation for this curve.

2 Sketch the curve whose parametric equations are $x = 2t, y = \dfrac{2}{t}$ and find a Cartesian equation for this curve.

3 Sketch the curve whose parametric equations are $x = t^2 + 2, y = t^3 + 4$.
Show that the curve also has equation $(y - 4)^2 = (x - 2)^3$.

4 For each of the following sets of parametric equations
  i) sketch the curve given by the equations;
  ii) obtain a Cartesian equation for the curve.

  a) $x = \cos\theta, y = \sin\theta$ **b)** $x - 3 + \cos\theta, y - 5 + \sin\theta$

  c) $x = 2\cos\theta, y = \sin\theta$ **d)** $x = 5\cos\theta, y = 2\sin\theta$

  e) $x = 5\cos\theta, y = 5\sin\theta$ **f)** $x = 2 + 3\cos\theta, y = 4 + 2\sin\theta$

Hence suggest parametric equations for the curves C_1 and C_2 shown in the diagram below.

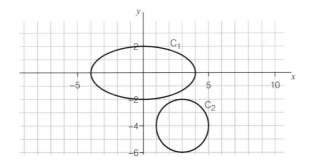

5 Sketch the curve given by
$$x = (1 + \cos\theta)\cos\theta, \quad y = (1 + \cos\theta)\sin\theta$$
and show that points on this curve satisfy the equation
$$(x^2 + y^2 - x)^2 = x^2 + y^2.$$

6 Sketch the curve given by
$$x = \cos\theta, \quad y = 2\cos2\theta$$
and prove that the curve is part but not all of the graph of $y = 4x^2 - 2$.

7 Prove that the curve whose parametric equations are
$$x = \ln t, \quad y = \ln(kt^n) \qquad (t > 0, k > 0)$$
is a straight line and express the gradient and y-intercept of the line in terms of k and n.

8 Sketch the curve given parametrically by $x = e^t, y = \dfrac{4}{1 + e^{2t}}$

and prove that the curve is part but not all of the graph of $y = \dfrac{4}{1 + x^2}$.

9 Sketch the curves given by
   a) $x = 5 \cos \theta + 2 \cos 2\theta$, $y = 5 \sin \theta + 2 \sin 2\theta$
   b) $x = 5 \cos \theta + 2 \cos 2\theta$, $y = 5 \sin \theta - 2 \sin 2\theta$

10 The diagram shows a wheel of radius 1 metre rolling along the x-axis. The point P is on the circumference of the wheel and is originally at the origin.

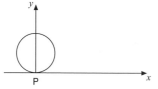 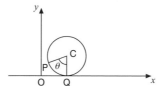

Since the wheel is rolling the distance OQ is equal to the length of the arc PQ.
   a) Write down the co-ordinates of C in the second diagram.
   b) Explain carefully why, in the second diagram, $y_P = 1 - \cos \theta$ and obtain an expression for x_P.
   c) Hence sketch the curve showing the locus of P as the wheel rolls along the x-axis.
   d) The diagram shows a flanged wheel which can roll along the x-axis.
   By making sensible approximations for the size of the wheel and the flange, investigate the locus of a point R on the flange.

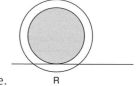

11 The diagram shows the graph whose parametric equations are

$$x = 5 \sin 3\theta, \quad y = 2 + 4 \sin \theta \qquad -\frac{\pi}{2} \leqslant \theta \leqslant \frac{\pi}{2}.$$

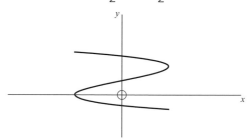

Find the values of the parameter at the points where the curve crosses the axes and deduce the co-ordinates of the points where the curve crosses the axes.

12 The diagram shows the graph whose parametric equations are

$$x = \sqrt{\theta} \cos \theta, \quad y = \sqrt{\theta} \sin \theta \qquad 0 \leqslant \theta \leqslant 2\pi.$$

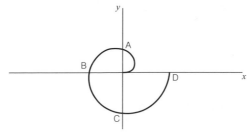

Find the exact co-ordinates of the points A, B, C and D.
Find also the points of intersection of the curve with the line $y = x$.

Frequently Occurring Parametric Representations of Curves

The table below lists some commonly occurring parametric equations of curves together with sketches of the curves and their Cartesian equations.

Throughout the table, t and θ denote parameters whilst a, b and c denote positive constants.

Parametric equations	Curve	Cartesian equations
$x = at^2$ $y = 2at$	 This is an example of a parabola.	$y = 2at \Rightarrow t = \dfrac{y}{2a}$ $x = at^2 \Rightarrow x = a\left(\dfrac{y}{2a}\right)^2$ $\Rightarrow x = \dfrac{ay^2}{4a^2}$ $\Rightarrow x = \dfrac{y^2}{4a}$ $\Rightarrow y^2 = 4ax$
$x = a\cos\theta$ $y = b\sin\theta$	 This is an example of an ellipse.	$x = a\cos\theta \Rightarrow \cos\theta = \dfrac{x}{a}$ $y = b\cos\theta \Rightarrow \sin\theta = \dfrac{y}{b}$ $\cos^2\theta + \sin^2\theta = 1$ $\Rightarrow \left(\dfrac{x}{a}\right)^2 + \left(\dfrac{y}{b}\right)^2 = 1$ $\Rightarrow \dfrac{x^2}{a^2} + \dfrac{y^2}{b^2} = 1$
$x = ct$ $y = \dfrac{c}{t}$	 This is an example of a rectangular hyperbola.	$x = ct \Rightarrow t = \dfrac{x}{c}$ $y = \dfrac{c}{t}$ $\Rightarrow y = \dfrac{c}{x/c}$ $\Rightarrow y = \dfrac{c}{1} \times \dfrac{c}{x}$ $\Rightarrow y = \dfrac{c^2}{x}$

The Gradient of a Curve Given Parametrically

The chain rule for differentiation tells us that

$$\frac{dy}{dt} = \frac{dy}{dx} \times \frac{dx}{dt}.$$

Rearranging this to make $\dfrac{dy}{dx}$ the subject gives

$$\frac{dy}{dx} = \frac{dy}{dt} \div \frac{dx}{dt}$$

and this is the result used for finding the gradient of parametric curves.

EXAMPLE 7

The point P(4, −8) lies on the curve $x = t^2$, $y = t^3$.

i) Find the value taken by the parameter t at the point P.
ii) Find the equation of the tangent to the curve at P.
iii) Find the co-ordinates of the point Q where this tangent meets the curve again.

i) Considering the x co-ordinate of P:

$$t^2 = 4 \Rightarrow t = \pm 2.$$

Considering the y co-ordinate of P:

$$t^3 = -8 \Rightarrow t = -2.$$

We conclude that, at P, $t = -2$.

> It is important that the value of t gives **both** co-ordinates correctly.

ii) $\dfrac{dy}{dx} = \dfrac{\dfrac{dy}{dt}}{\dfrac{dx}{dt}} = \dfrac{3t^2}{2t} = \dfrac{3t}{2}.$

Now P is the point where $t = -2$ so the gradient of the tangent at P is −3.
The tangent has gradient −3 and passes through (4, −8)

$$\Rightarrow \quad y - (-8) = -3(x - 4)$$
$$\Rightarrow \quad y = -3x + 4.$$

iii) A diagram helps to visualise the situation:

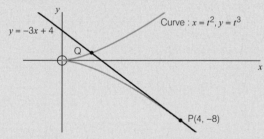

Q is the point where the curve $x = t^2$, $y = t^3$ meets the line $y = -3x + 4$:

$$\left. \begin{array}{r} x = t^2 \\ y = t^3 \\ y = -3x + 4 \end{array} \right\} \Rightarrow t^3 = -3t^2 + 4 \quad \Rightarrow \quad t^3 + 3t^2 - 4 = 0.$$

We know that the point P, where $t = -2$, is one point where the curve and the line meet so $(t + 2)$ must be a factor of the equation:

$$\Rightarrow \quad (t + 2)(t^2 + t - 2) = 0$$
$$\Rightarrow \quad (t + 2)(t + 2)(t - 1) = 0$$
$$\Rightarrow \quad t = 1 \quad \text{or} \quad -2.$$

Q is therefore the point where $t = 1$.
When $t = 1$, $x = 1^2 = 1$; $y = 1^3 = 1$, so Q is the point (1, 1).

EXAMPLE 8

a) A curve C has equation

$$x = e^{-\theta} \cos \theta, \; y = e^{-\theta} \sin \theta \qquad 0 \leqslant \theta \leqslant \pi.$$

Find $\dfrac{dy}{dx}$, giving your answer in terms of θ.

b) Find the points on the curve C where the gradient is zero.

c) Find also the points on the curve C where the tangent is parallel to the y-axis.

Illustrate your answers in a diagram.

a) The product rule applied to $x = e^{-\theta} \cos \theta$ gives

$$\begin{aligned}
\frac{dx}{d\theta} &= (-e^{-\theta})\cos \theta + e^{-\theta}(-\sin \theta) \\
&= -e^{-\theta} \cos \theta - e^{-\theta} \sin \theta \\
&= -e^{-\theta}(\cos \theta + \sin \theta).
\end{aligned}$$

Similarly, the product rule applied to $y = e^{-\theta} \sin \theta$ gives

$$\begin{aligned}
\frac{dy}{d\theta} &= (-e^{-\theta})\sin \theta + e^{-\theta}(\cos \theta) \\
&= -e^{-\theta} \sin \theta + e^{-\theta} \cos \theta \\
&= -e^{-\theta}(\sin \theta - \cos \theta).
\end{aligned}$$

Using $\dfrac{dy}{dx} = \dfrac{dy}{d\theta} \div \dfrac{dx}{d\theta}$ now gives

$$\frac{dy}{dx} = \frac{\dfrac{dy}{d\theta}}{\dfrac{dx}{d\theta}} = \frac{-e^{-\theta}(\sin \theta - \cos \theta)}{-e^{-\theta}(\cos \theta + \sin \theta)} = \frac{\sin \theta - \cos \theta}{\cos \theta + \sin \theta}.$$

b) If the gradient is zero then

$$\frac{\sin \theta - \cos \theta}{\cos \theta + \sin \theta} = 0$$

> Remember that if a fraction is zero then the numerator (i.e. the top) of the fraction must be zero.

$$\begin{aligned}
&\Longrightarrow \quad \sin \theta - \cos \theta = 0 \\
&\Longrightarrow \quad \sin \theta = \cos \theta \\
&\Longrightarrow \quad \tan \theta = 1 \\
&\Longrightarrow \quad \theta = \frac{\pi}{4}.
\end{aligned}$$

> This is the **only** solution of $\tan \theta = 1$ between 0 and π.

The point on the curve corresponding to $\theta = \dfrac{\pi}{4}$ is the point A with co-ordinates

$$\left(e^{-\frac{\pi}{4}} \frac{\sqrt{2}}{2}, \; e^{-\frac{\pi}{4}} \frac{\sqrt{2}}{2} \right).$$

EXAMPLE 8 (continued)

c) If the tangent is parallel to the y-axis then it must have an infinite gradient. The only way that the fraction $\dfrac{\sin\theta - \cos\theta}{\cos\theta + \sin\theta}$ can be infinite is if the denominator is zero.

$\Rightarrow \quad \cos\theta + \sin\theta = 0$
$\Rightarrow \quad \sin\theta = -\cos\theta$
$\Rightarrow \quad \tan\theta = -1$
$\Rightarrow \quad \theta = \dfrac{3\pi}{4}.$ ── This is the **only** solution of $\tan\theta = -1$ between 0 and π.

The point on the curve corresponding to $\theta = \dfrac{3\pi}{4}$ is the point B with co-ordinates

$$\left(-e^{-\frac{3\pi}{4}}\frac{\sqrt{2}}{2},\ e^{-\frac{3\pi}{4}}\frac{\sqrt{2}}{2}\right).$$

The curve $x = e^{-\theta}\cos\theta$, $y = e^{-\theta}\sin\theta$ for $0 \leqslant \theta \leqslant \pi$ is sketched below and the points A and B where the gradient is zero and where the tangent is parallel to the x-axis are marked on the sketch.

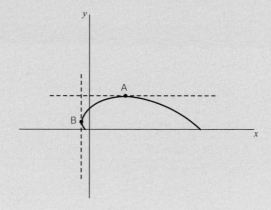

1 Find the equations of the tangent and of the normal to the curve

$x = t^2$, $y = 2t$ at the point $(9, 6)$.

2 Prove that the tangent to the curve

$x = 6t$, $y = \dfrac{6}{t}$ at the point $(12, 3)$

is $4y + x = 24$ and find the equation of the normal to the curve at this point.

3 If $x = t^3$, $y = t^2 - t$

a) find $\dfrac{\mathrm{d}y}{\mathrm{d}x}$ in terms of t;
b) find the co-ordinates of the turning point of the curve;
c) sketch the curve.

4 Find the equations of the tangent and normal to the curve

$$x = 12t, \ y = \frac{12}{t}$$

at the point A(24, 6).
Find the co-ordinates of the point B on the curve where the normal meets the curve again.
Illustrate your answer with a diagram.

5 Find the equation of the tangent to the curve

$$x = \frac{1}{t}, \ y = t^2$$

at the point (1, 1).
Find the co-ordinates of the point where this tangent meets the curve again.

6 Sketch the curve given by the parametric equation

$$x = 3t^2 + 2, \ y = -2t^3$$

Find the equation of the tangent to the curve at the point $(3p^2 + 2, -2p^3)$.
Show that this line is also the normal to the curve

$$x = t^2, \ y = 2t$$

at the point where $t = p$.

7 a) Sketch the curve given by the parametric equations

$$x = e^t, \ y = te^t.$$

b) Prove that the curve has just one stationary point and find its co-ordinates.

8 The diagram shows the curve defined by the parametric equations

$$x = \sin 2\theta + 2 \sin \theta, \ y = \cos 2\theta + 2 \cos \theta.$$

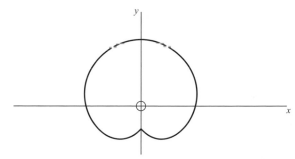

a) Find $\dfrac{dy}{dx}$, giving your answer in terms of θ.

b) Find the exact co-ordinates of the points of the curve where the tangent is parallel to either of the co-ordinate axes. Illustrate your answers on a copy of the diagram.

Implicit Differentiation

Sometimes the equation of a curve is given in terms of Cartesian co-ordinates but the relationship is so complex that it is either difficult or impossible to rewrite it in the usual explicit $y = f(x)$ format.

For example, the diagram below shows the curve of points satisfying the equation

$$x^4 + 4y^2 + 5x^2y^3 = 40.$$

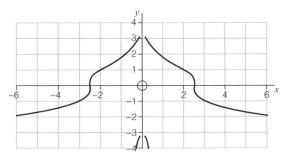

When a curve's Cartesian equation has been written in a form that is not of the $y = f(x)$ format we say that the curve's equation is an **implicit equation**.

When $x = 2$ and $y = 1$

$$x^4 + 4y^2 + 5x^2y^3 = 16 + 4 + 5 \times 4 \times 1 = 40$$

so the point $(2, 1)$ lies on the curve $x^4 + 4y^2 + 5x^2y^3 = 40$.

One might want to calculate the gradient of the curve at this point. Since the implicit equation $x^4 + 4y^2 + 5x^2y^3 = 40$ **cannot** be easily rewritten into the $y = f(x)$ format a method for differentiating implicit equations needs to be developed. The technique will be to differentiate the implicit equation term by term.

In order to differentiate the equation $x^4 + 4y^2 + 5x^2y^3 = 40$ term by term we must be able to handle expressions such as $\dfrac{d(y^2)}{dx}$ and $\dfrac{d(x^2y^3)}{dx}$.

Consider first the problem of finding $\dfrac{d(y^2)}{dx}$.

We know that $\dfrac{d(y^2)}{dy} = 2y$ but we do not know how to find the derivative **with respect to x** of y^2.

However, the chain rule states that

$$\frac{dz}{dx} = \frac{dz}{dy} \times \frac{dy}{dx}$$

and if y^2 is substituted for z this becomes

$$\frac{d(y^2)}{dx} = \frac{d(y^2)}{dy} \times \frac{dy}{dx} = 2y\frac{dy}{dx}.$$

Notation:

We write $\dfrac{d(y^2)}{dx}$ or $\dfrac{d}{dx}(y^2)$ to denote the derivative, with respect to x, of y^2.

Similarly, $\dfrac{d(t^2 e^{3x})}{dt}$ or $\dfrac{d}{dt}(t^2 e^{3x})$ denotes the derivative, with respect to t, of $(t^2 e^{3x})$.

Similarly, the derivative with respect to x of $x^2 y^3$ can be found using the product rule as well as the chain rule:

$$\frac{d(x^2 y^3)}{dx} = \frac{d(x^2 \times y^3)}{dx}$$

Product rule.

$$= \frac{d(x^2)}{dx} y^3 + x^2 \frac{d(y^3)}{dx}$$

$$= 2xy^3 + x^2 \frac{d(y^3)}{dy} \times \frac{dy}{dx}$$

Chain rule gives

$$\frac{d(y^3)}{dx} = \frac{d(y^3)}{dy} \times \frac{dy}{dx} = 3y^2 \frac{dy}{dx}.$$

$$= 2xy^3 + x^2 3y^2 \frac{dy}{dx}$$

$$= 2xy^3 + 3x^2 y^2 \frac{dy}{dx}.$$

EXERCISE 3

Find alternative expressions for

1 $\dfrac{d(y^4)}{dx}$

2 $\dfrac{d(e^{3y})}{dx}$

3 $\dfrac{d(\sin 3x)}{dt}$

4 $\dfrac{d(\cos 3\theta)}{dr}$

5 $\dfrac{d(\ln y)}{dx}$

6 $\dfrac{d(y^2 x^4)}{dx}$

7 $\dfrac{d(x^3 y^5)}{dx}$

8 $\dfrac{d(y^2 \ln x)}{dx}$

9 $\dfrac{d(x^3 e^{4y})}{dx}$

10 $\dfrac{d\left(\dfrac{y^2}{1+y}\right)}{dx}$

11 $\dfrac{d(r \cos \theta)}{d\theta}$

12 $\dfrac{d(e^{2y} \sin 3x)}{dx}$

Returning to the problem of finding the gradient of $x^4 + 4y^2 + 5x^2 y^3 = 40$ at the point $(2, 1)$, we proceed by differentiating each term of the curve equation with respect to x:

$$x^4 + 4y^2 + 5x^2 y^3 = 40$$

$$\Longrightarrow \quad \frac{d(x^4)}{dx} + \frac{d(4y^2)}{dx} + \frac{d(5x^2 y^3)}{dx} = 0$$

$$\Longrightarrow \quad 4x^3 + 4\frac{d(y^2)}{dx} + 5\frac{d(x^2 y^3)}{dx} = 0.$$

The results of the previous section now give

$$4x^3 + 4 \times 2y \frac{dy}{dx} + 5\left(2xy^3 + 3x^2 y^2 \frac{dy}{dx}\right) = 0$$

$$\Longrightarrow \quad 4x^3 + 10xy^3 + (8y + 15x^2 y^2)\frac{dy}{dx} = 0$$

$$\Longrightarrow \quad (8y + 15x^2 y^2)\frac{dy}{dx} = -(4x^3 + 10xy^3)$$

$$\Longrightarrow \quad \frac{dy}{dx} = -\frac{4x^3 + 10xy^3}{8y + 15x^2 y^2}.$$

So, at the point (2, 1)

$$\text{gradient} = -\frac{4 \times 2^3 + 10 \times 2 \times 1^3}{8 \times 1 + 15 \times 2^2 \times 1^2} = -\frac{52}{68} = -\frac{13}{17}.$$

EXAMPLE 9

Find the equation of the tangent to the curve $x^3 + 3xy^2 - y^3 = 13$ at the point (2, 1).

Differentiating $x^3 + 3xy^2 - y^3 = 13$ with respect to x gives

$$3x^2 + \frac{d(3xy^2)}{dx} - \frac{d(y^3)}{dx} = 0.$$

Implicit differentiation together with the product rule for the middle term gives

$$3x^2 + \left(3y^2 + 3x \times 2y \frac{dy}{dx}\right) - 3y^2 \frac{dy}{dx} = 0$$

$$\Rightarrow \quad 3x^2 + 3y^2 + (6xy - 3y^2)\frac{dy}{dx} = 0$$

$$\Rightarrow \quad \frac{dy}{dx} = \frac{3x^2 + 3y^2}{3y^2 - 6xy} = \frac{x^2 + y^2}{y^2 - 2xy}.$$

When $x = 2$ and $y = 1$

$$\text{gradient} = \frac{2^2 + 1^2}{1^2 - 2 \times 2 \times 1} = -\frac{5}{3}.$$

Since the tangent has gradient $-\frac{5}{3}$ and passes through the point (2, 1), it has equation

$$y - 1 = -\frac{5}{3}(x - 2)$$

$$\Rightarrow \quad 3y - 3 = -5x + 10$$

$$\Rightarrow \quad 3y + 5x = 13.$$

EXAMPLE 10

Find the points on the curve $16x^2 - y^2 = 144$ where the gradient is −5.

Differentiating $16x^2 - y^2 = 144$ with respect to x gives

$$32x - \frac{d(y^2)}{dx} = 0$$

$$\Rightarrow \quad 32x - 2y\frac{dy}{dx} = 0.$$

At points where the gradient is −5 this becomes

$$32x + 10y = 0.$$

If the point (x, y) lies on the curve $16x^2 - y^2 = 144$ and has gradient −5 then x and y must satisfy the simultaneous equations

$$\left.\begin{array}{l} 16x^2 - y^2 = 144 \\ 32x + 10y = 0 \end{array}\right\}.$$

EXAMPLE 10 (continued)

Rearranging the second equation to make y the subject gives

$$y = -\frac{16}{5}x$$

and substituting this into the first equation gives

$$16x^2 - \left(-\frac{16}{5}x\right)^2 = 144$$

$$\Rightarrow \quad 16x^2 - \frac{256}{25}x^2 = 144$$

$$\Rightarrow \quad 400x^2 - 256x^2 = 3600$$

$$\Rightarrow \quad 144x^2 = 3600$$

$$\Rightarrow \quad x^2 = 25$$

$$\Rightarrow \quad x = \pm 5.$$

Using $y = -\frac{16}{5}x$

when $x = 5$, $y = -16$ and when $x = -5$, $y = 16$.

The points on the curve $16x^2 - y^2 = 144$ where the gradient is -5 are $(5, -16)$ and $(-5, 16)$.

EXERCISE 4

1 Find $\dfrac{dy}{dx}$ in terms of x and y if

a) $y^3 - x^3 = 1$

b) $y^3 = 3(x - 2)$

c) $2x^2 - xy + y^2 + 3y - 4 = 0$

d) $\dfrac{1}{x} + \dfrac{1}{y^2} = 2$

c) $\sin x + \sin y - x - y$

f) $x^3 y^2 - x - 2y + 1$

2 Find the equations of the tangent and normal to the curve $4x^2 - 9y^2 = 64$ at the point $(5, -2)$.

3 a) Find $\dfrac{d\left(\dfrac{x}{1+x}\right)}{dx}$.

b) Find the equations of the tangent and normal to the curve $\dfrac{4x}{1+x} + \dfrac{5y}{1+y} = 6$ at the point $(1, 4)$.

4 Find the equation of the tangent to the curve $\sqrt{x} + \sqrt{y} = 5$ at the point $A(a^2, (5 - a)^2)$ where $0 < a < 5$.
The tangent meets the x-axis at P and the y-axis at Q. Show that the value of OP + OQ is independent of the value of a.

5 A curve is given by the equation $xy(x + y) = 16$.

    a) Prove that $\dfrac{dy}{dx} = -\dfrac{y(2x + y)}{x(x + 2y)}$.

    b) Find the point on the curve where the gradient is -1.

6 Find the co-ordinates of the points on the curve $x^2 + 4xy + y^2 = 132$ where the gradient is $-\dfrac{2}{3}$.

7 Prove that the tangent to the curve $x^3y^2 + 2xy^4 = 3$ at the point $(1, 1)$ has equation $2y + x = 3$.

8 Prove that the gradient of the curve $x^k + y^k = 1$, where k is a positive constant, is given by $\dfrac{dy}{dx} = -x^{k-1}y^{1-k}$.

Hence describe the tangent to the curve $x^k + y^k = 1$ at the point $(1, 0)$ if

    i) $0 < k < 1$ **ii)** $k > 1$

Describe also the tangent to the curve $x^k + y^k = 1$ at the point $(0, 1)$ if

    i) $0 < k < 1$ **ii)** $k > 1$

Sketch the curves

    a) $x^4 + y^4 = 1$ **b)** $x^3 + y^3 = 1$ **c)** $\sqrt{x} + \sqrt{y} = 1$

Having studied this chapter you should

- understand parametric equations for curves
- know that the gradient of a curve given parametrically in terms of a parameter t can be found using the formula $\dfrac{dy}{dx} = \dfrac{dy}{dt} \div \dfrac{dx}{dt}$
- be able to use implicit differentiation to find the gradient of a curve given by an implicit equation

REVISION EXERCISE

1 A curve C is defined by the parametric equations

$$x = t^3, \ y = t^2 + 2t .$$

    a) Express $\dfrac{dy}{dx}$ in terms of t.

    b) Find the equation of the tangent to the curve at the point where $t = -2$.

2 **i)** For the curve $2x^2 + xy + y^2 = 14$, find $\dfrac{dy}{dx}$.

    ii) Deduce that there are two points on the curve $2x^2 + xy + y^2 = 14$ at which the tangents are parallel to the x-axis, and find their co-ordinates.

(OCR Jun 2002 P3)

3   A curve is given by the parametric equations

$$x = t^2 - 1, \; y = t^3 - 12t.$$

a)   Find the co-ordinates of the points where the curve crosses the x-axis.

b)   Find the co-ordinates of the points where the curve crosses the y-axis.

c)   Find $\dfrac{dy}{dx}$ in terms of t and hence find the stationary points of the curve.

d)   Sketch the curve.

4   The parametric equations of a curve are $x = t^2$, $y = 2t$.

i)   Prove that the equation of the tangent at the point with parameter t is

$$ty = x + t^2.$$

ii)   The tangent at the point P(16, 8) meets the tangent at the point Q(9, −6) at the point R. Find the co-ordinates of R.

iii)   Prove that the line $2y - 4x - 1 = 0$ is a tangent to the curve, and find the parameter at the point of contact.

(OCR Jan 2002 P3)

5   Find the gradient of the curve $x^5 + 2y^5 = 30$ at the point where $x = 2$.

6   The parametric equations of a curve are

$$x = t + \frac{2}{t}, \; y = t - \frac{2}{t}$$

where $t \neq 0$.

i)   Show that $t = \dfrac{1}{2}(x + y)$ and hence write down the value of t corresponding to the point (3, 1) on the curve.

ii)   Find the gradient of the curve at the point (3, 1).

(OCR Jun 2004 P3)

7   Find the equation of the tangent to the curve

$$x^3 y^2 + 3x + 5y = 23$$

at the point (3, 1).

8   The equation of a curve is $3x^2 + y^2 = 2xy + 8x - 2$.

a)   Prove that $\dfrac{dy}{dx} = \dfrac{y + 4 - 3x}{y - x}$.

b)   Find the co-ordinates of the points on the curve where the gradient is 2.

9   The parametric equations of a curve are

$$x = \sin^3 \theta, \; y = 18 \sin \theta + \sqrt{27} \cos^2 \theta \quad \text{for} \quad 0 < \theta < \frac{1}{2}\pi.$$

i)   Find $\dfrac{dy}{dx}$ in terms of θ.

ii)   The tangent to the curve at the point P has gradient 4. Find the co-ordinates of P in exact form.

(OCR Jun 2000 P3)

10 The diagram shows the graph of the curve with parametric equations

$$x = t \sin t, \ y = 1 + \cos 2t \qquad 0 \leqslant t \leqslant 2\pi.$$

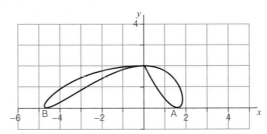

a) Find the values of t at the points A and B where the curve meets the x axis. Write down the exact co-ordinates of these points.

b) Find the values of t for which the curve passes through the point (0, 2).

c) Prove that, at the point where $t = \dfrac{\pi}{4}$ the curve has gradient $-\dfrac{8\sqrt{2}}{\pi + 4}$.

11 Find the gradient of the curve $x^2 + xy + 4y^2 = 10$ at the point (2, 1).
Find also the co-ordinates of the points at which the tangent to the curve is parallel to the y-axis.

12 A curve is given parametrically by the equations

$$x = 2t, \ y = \frac{1}{t^2}.$$

a) Sketch the curve.

b) Prove that the equation of the tangent to the curve at the point $P\left(2p, \dfrac{1}{p^2}\right)$ is

$$p^3 y + x = 3p$$

and write down the gradient of the normal to the curve at P.

c) This tangent meets the curve again at the point Q. Find the co-ordinates of Q.

d) If the tangent at P is the normal to the curve at Q show that $p = \pm\sqrt{2}$.

8 Differential Equations

The purpose of this chapter is to enable you to

- solve differential equations using the method of separation of variables

- formulate problems involving rates of change as differential equations, solve the differential equation and interpret the solution

In module C2 we met the simplest possible examples of differential equations. These were equations of the form

$$\frac{dy}{dx} = f(x).$$

These equations could be solved immediately by integration.

EXAMPLE 1

Solve the differential equation

$$\frac{dy}{dx} = \frac{x}{x^2 + 1} \qquad x = 0, \ y = 2.$$

Integrating each side of the equation

$$\frac{dy}{dx} = \frac{x}{x^2 + 1}$$

gives

$$y = \int \frac{x}{x^2 + 1} \, dx$$

$$= \frac{1}{2} \int \frac{2x}{x^2 + 1} \, dx$$

$$= \frac{1}{2} \ln(x^2 + 1) + c.$$

> The derivative of the denominator is $2x$ so we can use the result
>
> $$\int \frac{f'(x)}{f(x)} \, dx = \ln(|f(x)|) + c$$
>
> to simplify this integral.

The value of the integration constant can be found by using the initial conditions, $x = 0, \ y = 2$:

$$2 = \frac{1}{2} \ln(0^2 + 1) + c$$

$$\implies \quad c = 2.$$

The solution to the differential equation is

$$y = \frac{1}{2} \ln(x^2 + 1) + 2.$$

Solve the differential equations

1 $\dfrac{dx}{dt} = 4e^{2t}$ $\qquad t = 0, x = 3$

2 $\dfrac{dx}{dt} = 6 \sin 2t + 4 \cos 2t$ $\qquad t = \pi, x = 5$

3 $\dfrac{dx}{dt} = te^t$ $\qquad t = 0, x = -5$

4 $\dfrac{dx}{dt} = \dfrac{6 \cos 3t}{1 + \sin 2t}$ $\qquad t = \dfrac{\pi}{2}, x = 4$

5 $\dfrac{dx}{dt} = \dfrac{t^2}{(t+1)}$ $\qquad t = 0, x = 3$

Separating the Variables

Consider the equation $\dfrac{dy}{dx} = \dfrac{x^2}{y^4}$ $\qquad x = 0, y = 1$.

If we try to integrate this equation as it stands we obtain $y = \displaystyle\int \dfrac{x^2}{y^4} \, dx$ and the integral cannot be evaluated since it involves both x and y and we do not yet know how y is related to x.

This equation can be solved by the **method of separation of variables**. The first stage is to algebraically rearrange the equation so that all the y's appear on the same side of the equation as the $\dfrac{dy}{dx}$ whilst all the x's appear on the other side of the equation. This is called 'separating the variables'.

In the equation

$$\dfrac{dy}{dx} = \dfrac{x^2}{y^4} \qquad x = 0, y = 1$$

the variables can be separated by multiplying each side of the equation by y^4 to obtain

$$y^4 \dfrac{dy}{dx} = x^2. \qquad\qquad [1]$$

From the work on implicit differentiation in the last chapter, we know that

$$\dfrac{d}{dx}\left(\dfrac{1}{5}y^5\right) = \dfrac{d}{dy}\left(\dfrac{1}{5}y^5\right) \times \dfrac{dy}{dx} = y^4 \dfrac{dy}{dx}$$

so equation [1] can be rewritten as

$$\dfrac{d}{dx}\left(\dfrac{1}{5}y^5\right) = x^2$$

and integrating each side with respect to x gives

$$\dfrac{1}{5}y^5 = \int x^2 \, dx$$

$$\Rightarrow \quad \dfrac{1}{5}y^5 = \dfrac{1}{3}x^3 + c \qquad \text{where } c \text{ is the constant of integration.} \qquad [2]$$

This is the **general solution** of the differential equation $\dfrac{dy}{dx} = \dfrac{x^2}{y^4}$. The general solution of a differential equation will always contain an unknown constant.

The **initial conditions** or **boundary conditions** $x = 0$, $y = 1$ allow the value of c to be determined.

We know that $\dfrac{1}{5}y^5 = \dfrac{1}{3}x^3 + c$ and that when $x = 0$, $y = 1$

$\Rightarrow \qquad \dfrac{1}{5} = 0 + c$

$\Rightarrow \qquad c = \dfrac{1}{5}$

$\Rightarrow \qquad \dfrac{1}{5}y^5 = \dfrac{x^3}{3} + \dfrac{1}{5}$

$\Rightarrow \qquad y^5 = \dfrac{5}{3}x^3 + 1$

$\Rightarrow \qquad y = \sqrt[5]{\dfrac{5}{3}x^3 + 1}.$

In practice the movement from equation [1] to equation [2] is usually achieved by a slightly different argument:

$$y^4 \dfrac{dy}{dx} = x^2. \tag{1}$$

Integrating each side with respect to x gives

> This cancelling out of the dx's can be justified either by the implicit differentiation argument of the initial solution or by recalling the technique of integration by substitution which allows
>
> $\displaystyle \int \ldots \dfrac{dy}{dx}\, dx$ to be rewritten as $\displaystyle \int \ldots dy.$

$$\int y^4 \dfrac{dy}{dx}\, dx = \int x^2\, dx$$

$\Rightarrow \qquad \displaystyle \int y^4 \, dy = \int x^2 \, dx$

$\Rightarrow \qquad \dfrac{1}{5}y^5 = \dfrac{1}{3}x^3 + c \quad$ where c is the constant of integration. $\tag{2}$

The solution of the differential equation is completed, as before, by determining the value of c.

EXAMPLE 2

Find the general solution of the differential equation $\dfrac{dy}{dx} - 3x^2y^2 = 0$.

$\dfrac{dy}{dx} - 3x^2y^2 = 0$

$\Rightarrow \qquad \dfrac{dy}{dx} = 3x^2y^2$

$\Rightarrow \qquad \dfrac{1}{y^2}\dfrac{dy}{dx} = 3x^2.$

> In this case we are not told a point that the curve passes through. The value of the integration constant cannot therefore be found and our final solution will contain an unknown constant. This final solution is the **general solution** of the differential equation.

EXAMPLE 2 (continued)

Integrating each side with respect to x gives

$$\int \frac{1}{y^2} \frac{dy}{dx} \, dx = \int 3x^2 \, dx$$

$$\Rightarrow \quad \int \frac{1}{y^2} \, dy = \int 3x^2 \, dx$$

$$\Rightarrow \quad -\frac{1}{y} = x^3 + c$$

$$\Rightarrow \quad y = \frac{-1}{x^3 + c}.$$

EXAMPLE 3

Solve the differential equation $\dfrac{dy}{dx} = -\dfrac{x}{y}$ $\qquad x = 0, \ y = 5.$

$$\frac{dy}{dx} = -\frac{x}{y}$$

$$\Rightarrow \quad y\frac{dy}{dx} = -x$$

$$\Rightarrow \quad \int y\frac{dy}{dx} \, dx = \int -x \, dx$$

$$\Rightarrow \quad \int y \, dy = \int -x \, dx$$

$$\Rightarrow \quad \frac{1}{2}y^2 = -\frac{1}{2}x^2 + c.$$

We know that when $x = 0, \ y = 5$ so $12.5 = 0 + c \Rightarrow c = 12.5$

$$\Rightarrow \quad \frac{1}{2}y^2 = -\frac{1}{2}x^2 + 12.5$$

$$\Rightarrow \quad y^2 = 25 - x^2$$

$$\Rightarrow \quad y = \pm \sqrt{25 - x^2}$$

but we know that when $x = 0, \ y = 5$ so it must be the **positive** solution

$$\Rightarrow \quad y = \sqrt{25 - x^2}.$$

The key steps in solving a differential equation by the method of separation of the variables are to

- rewrite the equation into the '$q(y) \dfrac{dy}{dx} = p(x)$' format

- integrate each side of the equation with respect to x remembering that $\int q(y) \dfrac{dy}{dx} \, dx$

  can be rewritten as $\int q(y) \, dy$ and remembering that a constant of integration is needed

EXAMPLE 4

If $\dfrac{dy}{dt} = 0.5(100 - y)$ and $t = 0$, $y = 10$,

show by integration that

$$y = 100 - Ae^{-0.5t}$$

where A is a constant whose value should be determined.

Hence find the value of t when $y = 95$.

$$\frac{dy}{dt} = 0.5(100 - y).$$

Separating the variables

$$\Rightarrow \quad \frac{1}{100 - y}\frac{dy}{dt} = 0.5.$$

Integrating with respect to t

$$\Rightarrow \quad \int \frac{1}{100 - y}\frac{dy}{dt}\,dt = \int 0.5\,dt$$

$$\Rightarrow \quad \int \frac{1}{100 - y}\,dy = \int 0.5\,dt$$

$$\Rightarrow \quad -\ln|100 - y| = 0.5t + c.$$

When $t = 0$, $y = 10$ so $100 - y$ is positive which means that the modulus signs can be dropped:

$$\Rightarrow \quad -\ln(100 - y) = 0.5t + c$$

$$\Rightarrow \quad \ln(100 - y) = -0.5t - c$$

$$\Rightarrow \quad 100 - y = e^{-0.5t - c}$$

$$\Rightarrow \quad 100 - y = e^{-0.5t}e^{-c}$$

$$\Rightarrow \quad 100 - y = Ae^{-0.5t}$$

$$\Rightarrow \quad y = 100 - Ae^{-0.5t}.$$

> Remember $\ln x = p \Rightarrow x = e^p$.

> Take care with your algebra here: it is very easy to make a mistake.

> c is a constant so e^{-c} is also a constant, denoted here by A.

Using the initial conditions to find A:

$$10 = 100 - Ae^0 \quad \Rightarrow \quad 10 = 100 - A \quad \Rightarrow \quad A = 90$$

$$\Rightarrow \quad y = 100 - 90e^{-0.5t}.$$

If $y = 90$ we have

$$95 = 100 - 90e^{-0.5t}$$

$$\Rightarrow \quad 5 = 90e^{-0.5t}$$

$$\Rightarrow \quad \frac{1}{18} = e^{-0.5t}$$

$$\Rightarrow \quad \ln\left(\frac{1}{18}\right) = -0.5t$$

$$\Rightarrow \quad t = -2\ln\left(\frac{1}{18}\right) = 5.78 \quad \text{(to 2 d.p.)}$$

Solve the following differential equations:

1 $\dfrac{dy}{dx} = \dfrac{x}{y^2}$ $\quad\quad x = 1, y = 1$

2 $\dfrac{dy}{dx} = 2xy$ $\quad\quad x = 0, y = 1$

3 $\dfrac{dy}{dx} = \dfrac{x}{e^y}$ $\quad\quad x = 1, y = 0$

4 $\dfrac{dy}{dx} = \dfrac{1}{y^2}$ $\quad\quad x = 1, y = 1$

5 $\dfrac{dy}{dx} = 3y$ $\quad\quad x = 0, y = 2$

6 $\dfrac{dx}{dt} = 0.2(x - 20)$ $\quad\quad t = 0, x = 25$

7 $\dfrac{dy}{dx} = -\dfrac{2x}{y}$ $\quad\quad x = 6, y = 0$

8 $\dfrac{dy}{dx} = \dfrac{2y}{x}$ $\quad\quad x = 2, y = 4$

9 At any point on the curve C the gradient is twice the y co-ordinate of the point. The curve C passes through (0, 8). Find the equation of the curve.

10 A curve C has gradient $4\sqrt{y}(x + \sin 2x)$ at the point (x, y) and passes through the point (π, π^4). Find the equation of the curve.

Further Examples of Differential Equations

The integrations involved in solving a differential equation may well require the use of techniques of integration that have been met earlier in the module.

EXAMPLE 5

Solve the equation

$$e^{x+y}\dfrac{dy}{dx} = x \quad\quad x = -1, y = 0.$$

SOLUTION

$e^{x+y}\dfrac{dy}{dx} = x$

$\Rightarrow \quad e^x e^y \dfrac{dy}{dx} = x$

> To separate the variables, use the law of indices:
> $e^{x+y} = e^x e^y.$

$\Rightarrow \quad e^y \dfrac{dy}{dx} = xe^{-x}$

$\Rightarrow \quad \displaystyle\int e^y \dfrac{dy}{dx}\, dx = \int xe^{-x}\, dx$

$\Rightarrow \quad \displaystyle\int e^y\, dx = \int xe^{-x}\, dx.$

Integration by parts is required to determine $\displaystyle\int xe^{-x}\, dx.$

EXAMPLE 5 (continued)

If $\quad u = x \quad$ and $\quad \dfrac{dv}{dx} = e^{-x}$

then $\quad \dfrac{du}{dx} = 1 \quad$ and $\quad v = \displaystyle\int e^{-x} \, dx = -e^{-x}.$

So the integration by parts formula

$$\int u \frac{dv}{dx} \, dx = uv - \int \frac{du}{dx} \, v \, dx$$

gives

$$\int xe^{-x} \, dx = x(-e^{-x}) - \int 1(-e^{-x}) \, dx$$

$$\Longrightarrow \quad \int xe^{-x} \, dx = -xe^{-x} + \int e^{-x} \, dx$$

$$\Longrightarrow \quad \int xe^{-x} \, dx = -xe^{-x} - e^{-x} + c.$$

Returning to the solution of the differential equation, we now have

$$e^{y} = -xe^{-x} - e^{-x} + c.$$

We know that when $x = -1$, $y = 0 \quad \Longrightarrow \quad 1 = e^{1} - e^{1} + c \Longrightarrow c = 1$

$$\Longrightarrow \quad e^{y} = -xe^{-x} - e^{-x} + 1.$$

The solution could be left like this, or y could be made the subject of the solution:

$$\Longrightarrow \quad y = \ln(1 - e^{-x} - xe^{-x}).$$

EXAMPLE 6

Find the general solution of the equation

$$2(x^2 + 1)^2 \frac{dy}{dx} = x(y + 1).$$

SOLUTION

$$2(x^2 + 1)^2 \frac{dy}{dx} = x(y + 1)$$

$$\Longrightarrow \quad \frac{2}{y+1} \frac{dy}{dx} = \frac{x}{(x^2 + 1)^2}$$

$$\Longrightarrow \quad \int \frac{2}{y+1} \frac{dy}{dx} \, dx = \int \frac{x}{(x^2 + 1)^2} \, dx$$

$$\Longrightarrow \quad \int \frac{2}{y+1} \, dy = \int \frac{x}{(x^2 + 1)^2} \, dx.$$

The y-integral may be written down using an application of the result

$$\int \frac{f'(x)}{f(x)} \, dx = \ln(|f(x)|) + c$$

EXAMPLE 6 (continued)

to obtain

$$\int \frac{2}{y+1}\, dy = 2 \int \frac{1}{y+1}\, dy$$
$$= 2 \ln |y+1|.$$

The x-integral requires a substitution.

Putting $u = x^2 + 1$

gives

$$\frac{du}{dx} = 2x \implies `dx = \frac{1}{2x}\, du'.$$

So

$$\int \frac{x}{(x^2+1)^2}\, dx = \int \frac{x}{u^2}\frac{1}{2x}\, du$$

$$\implies \int \frac{x}{(x^2+1)^2}\, dx = \int \frac{1}{2u^2}\, du$$

$$\implies \int \frac{x}{(x^2+1)^2}\, dx = -\frac{1}{2u} + c$$

$$\implies \int \frac{x}{(x^2+1)^2}\, dx = -\frac{1}{2(x^2+1)} + c.$$

So

$$\int \frac{2}{y+1}\, dy = \int \frac{x}{(x^2+1)^2}\, dx$$

$$\implies 2 \ln|y+1| = \frac{-1}{2(x^2+1)} + c.$$

The general solution can be left in this form, or, with great care, the equation can be rewritten to make y the subject:

$$\ln|y+1| = \frac{-1}{4(x^2+1)} + k$$

> Notice that an arbitrary constant divided by 2 is still an arbitrary constant.

$$\implies |y+1| = \exp\left(\frac{-1}{4(x^2+1)} + k\right)$$

> $\exp(t)$ is just a different way of writing e^t.

$$\implies |y+1| = \exp\left(\frac{-1}{4(x^2+1)}\right)\exp(k)$$

> Using $e^{p+q} = e^p e^q$.

$$\implies |y+1| = A\exp\left(\frac{-1}{4(x^2+1)}\right)$$

> k is an arbitrary constant so $\exp(k) = e^k$ is also an arbitrary constant which we will call A.

$$\implies y+1 = \pm A\exp\left(\frac{-1}{4(x^2+1)}\right)$$

$$\implies y = -1 \pm A\exp\left(\frac{-1}{4(x^2+1)}\right).$$

> Notice how careful we must be in removing the logarithms and modulus function in this example.

EXERCISE 3

Solve the differential equations

1 $\dfrac{dy}{dx} = \dfrac{xy}{x^2+1}$ $x = 0, y = 1$

2 $\dfrac{dy}{dx} = \dfrac{1+y}{1+x}$ $x = 0, y = 3$

3 $2x\dfrac{dy}{dx} = (y-1)^2$

4 The gradient of a curve at the point (x, y) is $-\dfrac{8y^2}{(x-1)^3}$ and the curve passes through the

point $\left(2, -\dfrac{1}{3}\right)$. By solving a differential equation, prove that the equation of the curve is

$$y = \dfrac{x^2 - 2x + 1}{x^2 - 2x - 3}.$$

5 By solving the differential equation

$$(1 - e^x)\tan y\,\dfrac{dy}{dx} = e^x$$

find an expression for $\cos y$ in terms of x and a constant.

6 i) Express $\dfrac{1}{(x-2)(8-x)}$ as a sum of partial fractions.

ii) Hence find the solution of the differential equation

$$\dfrac{dy}{dx} = \dfrac{6y}{(x-2)(8-x)}\qquad (2 < x < 8)$$

given that $y = 4$ when $x = 5$.

7 a) Find $\displaystyle\int \dfrac{x}{\sqrt{x^2-9}}\,dx.$

b) Find the solution of the differential equation

$$\sqrt{x^2-9}\,\dfrac{dy}{dx} = 2x\sqrt{y}\qquad (x > 3)$$

for which $y = 16$ when $x = 5$, expressing y in terms of x.

8 a) Find $\displaystyle\int \tan^3 x \sec^2 x\,dx.$

b) Solve the differential equation

$$\cot^3 x\,\dfrac{dy}{dx} = -8y^2 \sec^2 x \qquad x = \dfrac{\pi}{4}, y = 1$$

and find the exact value of y when $x = \dfrac{\pi}{3}$.

9 **a)** Find $\int xe^{-3x}\,dx$.

b) Solve the differential equation

$$e^{3x}\frac{dy}{dx} = x\cos^2 y \qquad x = 0,\ y = \frac{\pi}{4}.$$

Using Differential Equations to Model Real Situations

Many scientific and economic phenomena can be expressed mathematically in terms of differential equations. The solution of these differential equations enables mathematical models to be constructed and predictions to be made.

The solution of such problems employs the modelling cycle which is often used to apply mathematical techniques to real problems. If you have studied module M1 you will already have encountered this modelling cycle.

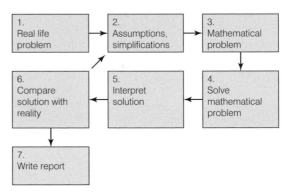

To illustrate the use of the modelling cycle with differential equations, consider the following examples.

EXAMPLE 7

A colony of bacteria is established on Agar jelly in a Petri dish.
At 10:00 one day a scientist notices that the colony occupies 2 cm² and at 11:30 he notices that the colony occupies 5 cm².
What area will the colony occupy at 15:00?

1 Real life problem: as stated in the question.

2 Simplifications and assumptions: it seems reasonable to assume that

- the area occupied by the colony is directly proportional to the number of bacteria;
- the rate of increase of the colony size is proportional to the size of the colony.

Putting these two together, we assume that the rate of increase of the occupied area is directly proportional to the area.

EXAMPLE 7 (continued)

3 **Pose the mathematical problem**: if A denotes the area occupied by the colony at t hours after 10:00 then we want to find the value of A when $t = 5$ given that

$$\frac{\mathrm{d}A}{\mathrm{d}t} \propto A; \qquad t = 0, A = 2 \quad \text{and} \quad t = 1.5, A = 5.$$

4 **Solve the mathematical problem**

Now $\dfrac{\mathrm{d}A}{\mathrm{d}t} \propto A \Longrightarrow \dfrac{\mathrm{d}A}{\mathrm{d}t} = kA$

where k is a constant of proportionality

$$\Longrightarrow \quad \frac{1}{A}\frac{\mathrm{d}A}{\mathrm{d}t} = k.$$

Integrating with respect to t gives

$$\int \frac{1}{A}\frac{\mathrm{d}A}{\mathrm{d}t}\,\mathrm{d}t = \int k\,\mathrm{d}t$$

$$\Longrightarrow \quad \int \frac{1}{A}\,\mathrm{d}A = \int k\,\mathrm{d}t$$

$$\Longrightarrow \quad \ln A = kt + c$$

$$\Longrightarrow \quad A = \mathrm{e}^{kt+c} = \mathrm{e}^{kt}\mathrm{e}^{c} = \alpha \mathrm{e}^{kt} \quad \text{where } \alpha \text{ is an arbitrary constant.}$$

Now, when $t = 0$, $A = 2$

$$\Longrightarrow \quad 2 = \alpha\,\mathrm{e}^{0} \Longrightarrow \alpha = 2$$

$$\Longrightarrow \quad A = 2\mathrm{e}^{kt}.$$

Now when $t = 1.5$, $A = 5$

$$\Longrightarrow \quad 5 = 2\mathrm{e}^{1.5k}$$

$$\Longrightarrow \quad 2.5 = \mathrm{e}^{1.5k}$$

$$\Longrightarrow \quad \ln 2.5 = 1.5k$$

$$\Longrightarrow \quad k = \frac{\ln 2.5}{1.5} = 0.61086\ldots$$

so we have

$$A = 2\mathrm{e}^{0.61086\ldots\,t}$$

and, putting $t = 5$ we get $A \approx 42.4 \text{ cm}^2$.

5 **Interpret solution**: our solution predicts that the colony will occupy approximately 42 cm^2 at 15:00.

6 **Compare with reality**: the size of the Petri dish, together with the availability of nutrients, will certainly limit the growth of the colony and mean that exponential growth is not sustainable.

(An improved model is considered in Q4 of Exercise 4.)

EXAMPLE 8

a) Find $\displaystyle\int \frac{1}{(70-w)(75-w)}\,dw$.

A chemical reaction takes place in a solution containing three substances, U, V and W. In the reaction 0.6 grams of U and 0.4 grams of V combine to make 1 gram of W.

Initially the solution contains 42 grams of U and 30 grams of V and there is no substance W present.

After one hour the solution contains 5 grams of substance W.

Given that the rate of production of W is proportional to the product of the amounts of U and V present in the solution, predict the future increase of substance W in the solution.

In particular, if w denotes the mass of substance W in the solution t hours after the start of the reaction

b) prove that there is a constant α such that

$$\frac{dw}{dt} = \alpha(70-w)(75-w)$$

c) solve this differential equation and sketch a graph to illustrate the relationship between t and w. What happens to the value of w as $t \to \infty$?

S
O
L
U
T
I
O
N

a) The expression $\displaystyle\frac{1}{(70-w)(75-w)}$ can be split into partial fractions and this will enable it to be integrated.

If

$$\frac{1}{(70-w)(75-w)} = \frac{A}{70-w} + \frac{B}{75-w}$$

then multiplying through by $(70-w)(75-w)$ gives

$$1 = A(75-w) + B(70-w).$$

Putting

$$w = 70 \Rightarrow 1 = 5A + 0 \Rightarrow A = \frac{1}{5}$$

$$w = 75 \Rightarrow 1 = 0 - 5B \Rightarrow B = -\frac{1}{5}$$

Since $\displaystyle\int \frac{1}{70-w}\,dw = -\int \frac{-1}{70-w}\,dw = -\ln|70-w|$

and similarly for $\displaystyle\int \frac{1}{75-w}\,dw$.

so

$$\frac{1}{(70-w)(75-w)} = \frac{1}{5}\left(\frac{1}{70-w} - \frac{1}{75-w}\right).$$

$$\int \frac{1}{(70-w)(75-w)}\,dw = \frac{1}{5}\int \left(\frac{1}{70-w} - \frac{1}{75-w}\right)dw$$

$$= \frac{1}{5}\left(-\ln|70-w| + \ln|75-w|\right) + c.$$

$$= \frac{1}{5}\left(\ln|75-w| - \ln|70-w|\right) + c.$$

EXAMPLE 8 (continued)

The modelling cycle is illustrated by parts (b) and (c) of the example:

1 **Real life problem**: as stated in the question.

2 **Simplifications and assumptions**: in this case these have been stated in the question. The critical piece of information is the statement
   '*the rate of production of W is proportional to the product of the amounts of U and V present in the solution*'.

3 **Pose the mathematical problem**: if w denotes the mass of substance W in the solution at time t hours, then

   - $0.6w$ grams of substance U have been used in making the w grams of substance W. This means that there are still $42 - 0.6w$ grams of U left in the solution.
   - $0.4w$ grams of substance V have been used in making the w grams of substance W. This means that there are still $30 - 0.4w$ grams of V left in the solution.
   - We are told that

$$\frac{dw}{dt} \propto \text{mass of U present in solution} \times \text{mass of V present in solution}$$

$$\Rightarrow \quad \frac{dw}{dt} \propto (42 - 0.6w)(30 - 0.4w)$$

$$\Rightarrow \quad \frac{dw}{dt} = k(42 - 0.6w)(30 - 0.4w)$$

where k is a constant of proportionality.

This is not quite the required form for $\dfrac{dw}{dt}$. However, observing that

$(42 - 0.6w) = 0.6(70 - w)$ and $(30 - 0.4w) = 0.4(75 - w)$

$$\Rightarrow \quad \frac{dw}{dt} = k \times 0.6(70 - w) \times 0.4(75 - w)$$

$$\Rightarrow \quad \frac{dw}{dt} = \alpha(70 - w)(75 - w) \quad \text{where} \quad \alpha = 0.24k$$

which is the required form.

We want to solve the differential equation

$$\frac{dw}{dt} = \alpha(70 - w)(75 - w)$$

subject to the conditions $t = 0,\ w = 0;\ t = 1,\ w = 5.$

EXAMPLE 8 (continued)

4   **Solve the mathematical problem**

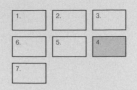

$$\frac{dw}{dt} = \alpha(70 - w)(75 - w)$$

$$\implies \frac{1}{(70 - w)(75 - w)} \frac{dw}{dt} = \alpha.$$

Integrating with respect to t gives

$$\int \frac{1}{(70 - w)(75 - w)} \frac{dw}{dt} \, dt = \int \alpha \, dt$$

$$\implies \int \frac{1}{(70 - w)(75 - w)} \, dw = \int \alpha \, dt.$$

Using part (a) of the question, we can write

$$\implies \frac{1}{5} \left(\ln|75 - w| - \ln|70 - w| \right) + c = \alpha t$$

where c is an integration constant.

Since w is growing from 0, $70 - w$ and $75 - w$ are both positive so the modulus signs can safely be dropped:

$$\implies \frac{1}{5} \left(\ln(75 - w) - \ln(75 - w) \right) + c = \alpha t$$

$$\implies \frac{1}{5} \ln\left(\frac{75 - w}{70 - w}\right) + c = \alpha t. \qquad [1] \qquad \boxed{\ln a - \ln b = \ln\left(\tfrac{a}{b}\right)}$$

We know that when $t = 0$, $w = 0$

$$\implies \frac{1}{5} \ln\left(\frac{75}{70}\right) + c = 0$$

$$\implies c = -\frac{1}{5} \ln\left(\frac{15}{14}\right).$$

We also know that when $t = 1$, $w = 5$. Equation [1] gives

$$\frac{1}{5} \ln\left(\frac{70}{65}\right) - \frac{1}{5} \ln\left(\frac{15}{14}\right) = \alpha$$

$$\implies \alpha = \frac{1}{5} \left(\ln\left(\frac{14}{13}\right) - \ln\left(\frac{15}{14}\right) \right) = \frac{1}{5} \ln\left(\frac{14}{13} \div \frac{15}{14}\right) = \frac{1}{5} \ln\left(\frac{196}{195}\right) = 0.001023....$$

Returning to equation [1], we have

$$\frac{1}{5} \ln\left(\frac{75 - w}{70 - w}\right) - \frac{1}{5} \ln\left(\frac{15}{14}\right) = \alpha t$$

$$\implies \ln\left(\frac{75 - w}{70 - w}\right) - \ln\left(\frac{15}{14}\right) = 5\alpha t$$

$$\implies \ln\left(\frac{75 - w}{70 - w} \div \frac{15}{14}\right) = 5\alpha t$$

EXAMPLE 8 (continued)

$$\Rightarrow \quad \ln\left(\frac{14(75-w)}{15(70-w)}\right) = 5\alpha t$$

$$\Rightarrow \quad \frac{14(75-w)}{15(70-w)} = e^{5\alpha t}$$

$$\Rightarrow \quad 14(75-w) = 15e^{5\alpha t}(70-w)$$

$$\Rightarrow \quad 1050 - 14w = 1050e^{5\alpha t} - 15e^{5\alpha t}w$$

$$\Rightarrow \quad 15e^{5\alpha t}w - 14w = 1050e^{5\alpha t} - 1050$$

$$\Rightarrow \quad w(15e^{5\alpha t} - 14) = 1050(e^{5\alpha t} - 1)$$

$$\Rightarrow \quad w = \frac{1050(e^{5\alpha t} - 1)}{(15e^{5\alpha t} - 14)} \quad \text{where} \quad \alpha = \frac{1}{5}\ln\left(\frac{196}{195}\right) = 0.001023....$$

5 Interpret solution: a graph of w against t would be an appropriate means of interpreting the solution.

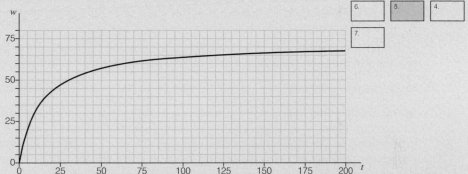

The amount of substance W grows quite quickly to begin with but levels out as time progresses and appears to have a maximum possible value of 70.

The limiting value of 70 grams can be explained by considering the formula for w:

$$w = \frac{1050(e^{5\alpha t} - 1)}{(15e^{5\alpha t} - 14)}$$

$$= \frac{1050(e^{5\alpha t} - 1)}{(15e^{5\alpha t} - 14)} \times \frac{e^{-5\alpha t}}{e^{-5\alpha t}}$$

$$= \frac{1050(1 - e^{-5\alpha t})}{(15 - 14e^{-5\alpha t})} .$$

As $t \rightarrow \infty$, $e^{-5\alpha t} \rightarrow 0$ so $w \rightarrow \frac{1050(1-0)}{(15-0)} = \frac{1050}{15} = 70.$

6 Compare with reality: if you study Chemistry you may be able to comment on the reliability of the assumptions and simplifications made in producing this model.

EXERCISE 4

1 The rate of decay, at any instant, of a radioactive substance is proportional to the amount, x grams, of the substance present at that instant.
At a certain time there is 100 grams of a radioactive material present. One hour later only 90 grams remain.
Let t denote the time, in hours, after the instant when $x = 100$.

a) Explain why there must be a positive constant k such that $\dfrac{dx}{dt} = -kx$.

b) By solving this differential equation and using the fact that $x = 100$ when $t = 0$, prove that $x = 100e^{-kt}$.

c) Find the value of k.

d) Find the amount remaining 6 hours from the start.

e) Find the time when only 5 grams of the substance remain.

2 During a spell of freezing weather, the ice on a pond has thickness x mm at time t hours after the start of freezing. At 3:00 p.m., after one hour of freezing weather, the ice is 2 mm thick and it is desired to predict when it will be 4 mm thick.

i) In a simple model, the rate of increase of x is assumed to be constant. For this model, express x in terms of t and hence determine when the ice will be 4 mm thick.

ii) In a more refined model, the rate of increase of x is taken to be proportional to $\dfrac{1}{x}$.

Set up a differential equation for x, involving a constant of proportionality k.
Solve the differential equation and hence show that the thickness of ice is proportional to the square root of the time elapsed from the start of freezing.
Determine the time at which the second model predicts the ice will be 4 mm thick.

iii) What assumptions about the weather underlie both models?

(OCR Nov 1995 P3)

3 Newton's law of cooling states that the rate of temperature loss of an object is directly proportional to the amount by which the object's temperature exceeds the temperature of the surroundings.

i) A cup of coffee whose temperature is 98 °C is put outside on a day when the temperature is 12 °C.
After 10 minutes the temperature of the coffee has fallen to 58 °C.
After t minutes the temperature of the coffee is $\theta°$.

a) Explain why there must be a positive constant k so that θ satisfies the equation
$\dfrac{d\theta}{dt} = -k(\theta - 12)$.

b) Solve this differential equation and find the value of k.

c) Sketch a graph to show how θ varies with t.

d) How long does it take for the temperature to fall below 30 °C?

ii) The normal body temperature of a living person is 98.4 °F. The body of a murder victim was discovered in a room at 22:00 one evening. The police doctor arrived at 22:30 and immediately took the temperature of the body, which was 94.6 °F. One hour later he again took the temperature and found that it was 93.4 °F. He noted that the temperature of the room was 70 °F.
Estimate the time of the murder.

4 A school supplies computers at cut prices to its 1500 students. Each student is allowed to purchase at most one computer. After t weeks, y students have taken up this offer. The number of students buying their computers per week is assumed to be one-tenth of the number of students who have yet to purchase a computer.

Taking y and t to be continuous variables, this situation can be modelled by the differential equation

$$\frac{dy}{dt} = \alpha(\beta - y)$$

where α and β are constants.

i) State the values of α and β.

ii) Solve the differential equation given that $y = 0$ when $t = 0$.

iii) According to this model, after how many weeks will 80% of the students have purchased a cut-price computer?

5 In an attempt to model cell division in a specimen, the rate of increase of the number of cells is assumed to be proportional to the number of cells, N, present at time t days. It is also assumed that N can be treated as a continuous variable.

i) Write down a differential equation for N involving a constant of proportionality and find the general solution, expressing N in terms of t.

ii) For one particular specimen, the number of cells is 2050 when $t = 0$ and 3075 when $t = 1$. Find a formula for N in terms of t.

In a revised model it is assumed that

$$\frac{dN}{dt} = 0.5(N - 470).$$

iii) Solve this differential equation given that $N = 2050$ when $t = 0$ and verify that this model predicts approximately the same value of N when $t = 1$ as was used in part (ii).

iv) Describe and compare the predictions of the two models over a longer period of time. Justify your comparison.

(OCR Nov 1999 P3)

6 The initial amount of mineral ore in a deep mine is taken to be 100 units. The amount of ore remaining after t years is x units. A model suggests that the rate of extraction of the ore is proportional to the square root of amount of ore remaining.

a) Write down a differential equation in $\frac{dx}{dt}$ to represent this model.

b) Show that

$$20 - 2\sqrt{x} = kt$$

where k is a positive constant.

c) After 10 years, 64 units of the original ore remain. Calculate the further number of years that will elapse, according to this model, until 16 units of the ore remain in the mine.

7 (Improved model for population growth.)

The model considered in example 7 on pages 132 and 133 for the area of the Petri dish occupied by the culture assumed that the rate of increase in the area occupied by the colony was proportional to the area already occupied by the colony. That is

$$\frac{dA}{dt} = kA.$$

We have seen this leads to exponential growth. This can obviously not be sustained since the Petri dish has finite area.

Now suppose that the area of the Petri dish is 100 cm^2 and that the rate of increase of the area is proportional to the product of the occupied area and the unoccupied area – this reflects that the rate of growth depends on **both** the number of bacteria already present **and** the space available for the colony to grow into.

a) Explain how this leads to the differential equation

$$\frac{dA}{dt} = pA(100 - A) \qquad t = 0, A = 2; t = 1.5, A = 5 \qquad\qquad [1]$$

where p is a positive constant.

b) Express $\dfrac{1}{A(100 - A)}$ as a sum of partial fractions and hence find

$$\int \frac{1}{A(100 - A)} \, dA.$$

c) By solving the differential equation [1], prove that

$$100pt = \ln\left(\frac{49A}{100 - A}\right)$$

and find the value of p.

d) Hence find A as a function of t.

e) Determine the value of A when $t = 5$. Compare this with the previous answer.

f) Sketch, on a single set of axes, the two possible graphs of A against t. Comment on your graphs.

8 At time $t = 0$, 200 rats escaped from a boat onto an island which until that time had no rat population. t years later the island had a population of N rats. Two mathematical models are suggested for the growth of N.

Model 1: $\dfrac{dN}{dt} = 0.2N.$

Model 2: $\dfrac{dN}{dt} = (0.2 + 0.1k \cos kt)N$ where $k = 2\pi.$

a) Find the population of rats predicted by model 1, 15 months after the initial escape.

b) Find the population of rats predicted by model 2, 15 months after the initial escape.

c) Use a graphical calculator or a computer graph-drawing package to produce a single diagram showing the growth of N, over a 5-year period, predicted by each of the models.

d) Describe and explain verbally the difference between the two models.

9 A solid sphere has uniform density. The mass of the sphere is denoted by m and the surface area of the sphere is denoted by A.

i) Explain why $A \propto m^{\frac{2}{3}}$.

A simplified model for the growth of the cells states that the rate of increase of the mass of the cell depends primarily on the cell's ability to take nutrients in across its cell walls from its surroundings. The rate of increase of the mass of the cell is therefore assumed to be proportional to the surface area of the cell.

ii) Assuming that the cells are spherical, show how this model leads to the differential equation

$$\frac{dm}{dt} = km^{\frac{2}{3}} \qquad \text{where } k \text{ is a positive constant.}$$

iii) Find the general solution of this differential equation.

iv) Given that, using appropriate units, $m = 1$ when $t = 0$ and $m = 1.5$ when $t = 2$, find the value of t when $m = 2$.

10 A new form of TV has recently been introduced into Britain. Currently just 1% of households have one of these new TVs. Let P denote the percentage of British households with one of these TVs in t years' time.

A model for the growth of P suggests that the rate of growth of P is the sum of two terms.

The first term takes account of the increase in P due to people with the new form of TV impressing their friends who do not have the new TVs. This term must take into account the number of people who have the new TVs and the number of people who do not have the TVs. The model assumes this term is directly proportional to the product of P and $(100 - P)$.

The second term takes account of the increase in P due to people without the new form of TV seeing the new TVs in the shops and in advertisements. The model assumes this term is proportional to $(100 - P)$.

a) Show that the model described leads to the differential equation

$$\frac{dP}{dt} = (\alpha P + \beta)(100 - P) \qquad t = 0, P = 1.$$

b) If the constants α and β are 0.005 and 0.1 respectively, show that the differential equation becomes

$$\frac{dP}{dt} = \frac{1}{200}(P + 20)(100 - P) \qquad t = 0, P = 1$$

and solve this equation.

Hence

i) find the time which must elapse from now until ownership of the TV reaches 75% of households;

ii) use a graphical calculator or computer software package to obtain a graph P against t.

Having studied this chapter you should know

- that a differential equation of the form $\dfrac{dy}{dx} = p(x)q(y)$ can be solved by first separating the variables to obtain $\dfrac{1}{q(y)}\dfrac{dy}{dx} = p(x)$ and then integrating each side with respect to x to obtain $\displaystyle\int \dfrac{1}{q(y)}\,dy = \int p(x)\,dx$

- how to formulate problems as differential equations and then solve the differential equations

REVISION EXERCISE

1 Solve the differential equation

$$(4 + x^2)\frac{dy}{dx} = 4x\sqrt{y} \qquad x = 0,\ y = 4(\ln 2)^2$$

giving your final answer in the form $y = f(x)$.

2 A model for the mass loss of a person following a specific diet suggests that daily rate of mass loss is approximately proportional to the mass of the person at the time.
On the day she starts following this diet, Jean had a mass of 90 kg. Let m denote Jean's mass t days after she started the diet.

i) Use the model to write down a differential equation linking m and t.

ii) Prove that $m = 90e^{-kt}$ where k is a positive constant.

A diet advisor informs Jean that after 30 days her mass should be 85 kg.

iii) Use this information to determine the value of the constant k.

iv) How long should Jean expect to have to follow this diet to reduce her mass to 65 kg?

Sketch a graph showing how m varies with t. Explain why this model for mass loss is unlikely to be valid if the diet is followed for long periods of time.

3 The variables x and t are linked by the differential equation

$$\frac{dx}{dt} = k(50 - x) \qquad t = 0,\ x = 0$$

where k is a positive constant and $x < 50$ for all $t \geqslant 0$.

a) Prove that $x = 50(1 - e^{-kt})$.

b) Given that $x = 10$ when $t = 5$, prove that $k = \dfrac{1}{5}\ln\left(\dfrac{5}{4}\right)$.

c) Sketch the graph of x against t and
hence find the values of t for which $x > 40$.

4 In the x–y plane, O is the origin, A is the point $(4, 0)$ and P is a general point (x, y). Write down the gradients of OP and AP.

A curve has the property that the gradient of the normal to the curve at the general point P(x, y) is equal to the product of the gradients of OP and AP.
Show that

$$\frac{dy}{dx} = \frac{4x - x^2}{y^2}.$$

Find the equation of the curve given that it passes through the point $(2, 3)$.

5 Given that

$$\frac{dy}{dx} = (x \ln x)y^{\frac{1}{2}}$$

and that $y = 1$ when $x = 1$, show that the value of y when $x = 3$ is $\left(\frac{9}{4}\ln 3\right)^2$.

(OCR Jun 2001 P3)

6

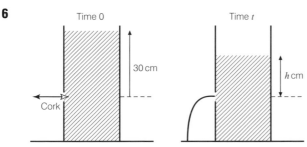

A cylindrical vessel has a cork in its side wall.
The vessel is filled with water so the water level is 30 cm above the cork.

At time 0, the cork is pulled out. t seconds later the water is h cm above the level of the hole.

A model for the flow of the water from the vessel states that the rate of decrease of h is proportional to the square root of h.

i) Write this information as a differential equation and solve the differential equation.

After 10 seconds the water level is observed to have dropped to 25 cm above the hole.

ii) Find the time that elapses from the instant when the cork is extracted from the vessel until the flow of water ceases.

iii) Sketch a graph to show how h varies with t.

7 A population grows at a rate proportional to the population at that time. Initially the population size is N_0 and after t years the population has size N.

a) Write down a differential equation relating N and t.

b) By solving the differential equation show that $t = \frac{1}{k}\ln\left(\frac{N}{N_0}\right)$, where k is a positive constant.

c) Given that $N = 2N_0$ when $t = 8$, calculate the time taken for the population size to become $5N_0$, giving your answer to three significant figures.

8 The co-ordinates of the points (x, y) of points on a curve C satisfy the differential equation

$$\frac{dy}{dx} = -\frac{x-1}{y+1}.$$

i) Find the general solution of this differential equation.

ii) The curve C passes through the origin. Deduce from your answer to part (i) that the equation of C may be written as

$$(x-1)^2 + (y+1)^2 = 2.$$

iii) Sketch the curve C.

(OCR Jun 2003 P3)

9 In a model of the depreciation of the value of a computer, it is assumed that the value $£C$, at age t months, decreases at a rate which is proportional to C. Using this model, write down a differential equation relating C and t. If the initial value of the computer is £1500, solve the differential equation and show that

$$C = 1500e^{-kt}$$

where k is a positive constant.

The value of the computer is expected to decrease to £1000 after 15 months. Calculate

i) the value, to the nearest pound, predicted by this model for the value of the computer after three years;

ii) the age of the computer, to the nearest month, when its value is £800.

10 Find the solution of the differential equation

$$\frac{dy}{dx} = \frac{x^4 - 1}{x^2 y^2}$$

given that $y = 2$ when $x = 2$.

Revise chapters 3, 5 and 7 before attempting this exercise.

1 Find the expansion, in ascending powers of x up to the term in x^3, of $\dfrac{1}{(1+2x)^2}$.
State the range of values of x for which the expansion is valid.

2 Find the first four terms of the expansion in ascending powers of x of $\sqrt{4-3x}$.
State the values of x for which the expansion is valid.

3 **a)** Simplify $\dfrac{x^3 - 64x}{2x^2 + 12x - 32}$.

b) Write $\dfrac{5}{x+2} + \dfrac{1}{x-1} - \dfrac{2}{(x-1)^2}$ as a single fraction.

4 Given that $|x| < \dfrac{1}{3}$, expand $(1-3x)^{\frac{1}{3}}$ in ascending powers of x, up to and including the term in x^2, simplifying the coefficients.

(OCR Jan 2004 P3)

5 **a)** Find the expansions, in ascending powers of x up to the x^3 term, of

i) $\dfrac{1}{1+2x}$ **ii)** $\dfrac{1}{1-5x}$

and state the values of x for which these expansions are valid.

b) Express $f(x) = \dfrac{5-11x}{(1+2x)(1-5x)}$ as a sum of two partial fractions.

c) Hence obtain the expansion, in ascending powers of x up to the x^3 term, of $f(x)$ and state the values of x for which these expansions are valid.

6 The diagram shows the curve with parametric equations
$$x = t^2 - 9, \; y = 4 - e^{-t}.$$

Find the values of t and the co-ordinates of the points where the curve crosses the co-ordinate axes.

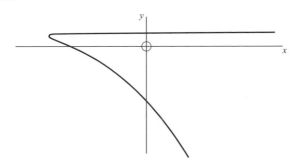

7 i) Find the expansion, up to the x^2 term, of

    a) $(1 + 2x)^{\frac{1}{2}}$ **b)** $(1 - 4x)^{-\frac{1}{2}}$

  ii) Hence find the expansion, up to the x^2 term, of $\sqrt{\dfrac{1 + 2x}{1 - 4x}}$.

 iii) By putting $x = 0.01$, use this expansion to estimate the value of $\sqrt{17}$.

8 Express $g(x) = \dfrac{x^2 + 3x + 3}{x(x + 1)(x + 2)}$ as a sum of partial fractions.

Hence evaluate $\displaystyle\int_1^2 g(x)\, dx$.

9 a) Express $g(x) = \dfrac{3x^2 + 10x + 9}{(x + 2)^2(x + 1)}$ as a sum of partial fractions.

  b) Hence prove that

$$\int_0^1 g(x)\, dx = \ln 6 - \frac{1}{6}.$$

10 a) In the expansion of $(1 + ax)^{\frac{1}{2}}$ the coefficient of x^3 is 32. Find the value of a.

  b) In the expansion of $(1 + 8x)^a$ the coefficient of x^2 is 24. Find the possible values of a.

11 A curve C is given parametrically by the equations

    $x = \sin t,\ y = \cos 2t$ $0 \leqslant t \leqslant 2\pi$.

  a) Show that all points on C are also points on the curve $y = 1 - 2x^2$.

  b) Sketch the graph of $y = 1 - 2x^2$.

  c) Explain briefly why there are points on the graph of $y = 1 - 2x^2$ which are not points of C.

  d) Indicate clearly on your diagram the points that are on the curve C.

12 Find values A, B, C and D so that $\dfrac{3x^3 + 2x^2 - 3x + 13}{2 + x} \equiv Ax^2 + Bx + C + \dfrac{D}{2 + x}$

and hence evaluate

$$\int_{-4}^{-3} \frac{3x^3 + 2x^2 - 3x + 13}{2 + x}\, dx.$$

13 i) Expand $(1 + 4x)^{\frac{1}{2}}$ in ascending powers of x, up to and including the term in x^2, simplifying the coefficients.

  ii) State the values of x for which the expansion is valid.

 iii) In the expansion of

    $(1 + kx)(1 + 4x)^{\frac{1}{2}}$

the coefficient of x is 7. Find the value of the constant k and the coefficient of x^2.

(OCR Jan 2002 P3)

14 Find the first three terms of the expansion in ascending powers of x of $\dfrac{1}{\sqrt{9 - 4x}}$ and state the values of x for which the expansion is valid.

Differentiation and Integration

Revise chapters 1, 2, 6 and 7 before attempting this exercise.

1 A curve is given parametrically by the equations

$$x = t(1 + t), \ y = t^2(1 + t).$$

i) Find $\dfrac{dy}{dx}$ in terms of t.

ii) Find the equation of the tangent to the curve at the point where $t = 2$, giving your answer in the form $ax + by + c = 0$.

iii) By first simplifying $\dfrac{y}{x}$, show that the curve has Cartesian equation

$$x^3 = xy + y^2.$$

(OCR Jun 2001 P3)

2 The equation of a curve is $x^3 + y^5 = 7$.
Find the equation of the normal to the curve at the point $(2, -1)$.

3 The equation of a curve is $y = xe^{-\frac{1}{2}x}$.
Find the co-ordinates of the turning point on the curve and determine whether it is a maximum or a minimum point.
Sketch the curve.
Prove that the area of the closed region bounded by the curve, the x- and y-axes and the line $x = \ln 4$ is $2 - 2 \ln 2$.

4 Find $\dfrac{dy}{dx}$ if

a) $y = (1 + \sin 3x)^3$ **b)** $y = \ln(3 + 2 \cos 4x)$.

5 By means of the substitution $u = \sin x$, find $\displaystyle\int \cos x \cos 2x \, dx$.

(OCR Jun 2000 P3)

6 Given that $y = \sin(x^3)$, find $\dfrac{d^2 y}{dx^2}$.

(OCR Jun 1997 P3)

7 Find

i) $\displaystyle\int x^5 \ln x \, dx$

ii) the exact value of $\displaystyle\int_0^{\frac{1}{4}} x(4x - 1)^4 \, dx$.

8 Find the equation of the normal to the curve $y = 4 \cos^2 2x$ at the point $\left(\dfrac{\pi}{6}, 1\right)$.

9 Use the substitution $u = 4 + x^2$ to show that $\displaystyle\int_0^1 \frac{x^3}{\sqrt{4+x^2}}\,dx = \frac{1}{3}(16 - 7\sqrt{5}).$

(OCR Feb 1997 P3)

10 **a)** Sketch the curve whose equation is given parametrically by the equations

$$x = 4t, \ y = t^2$$

and determine the Cartesian equation of this curve.

b) Prove that the equation of the normal to the curve at the point where $t = 3$ is

$$3y + 2x = 51.$$

c) The normal meets the curve again at the point Q. Find the co-ordinates of the point Q.

11 Find the gradient of the curve $x^2 + 4xy + y^2 = 25$ at the point where the curve meets the positive x-axis.

12 The diagram shows the curve C defined by $y = x^2 \ln x$ ($x > 0$), the line L with equation $y = ex$ and part of the line $x = 1$.

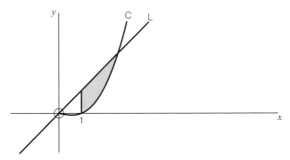

a) Determine the co-ordinates of the stationary point of C.

b) Show that C and L intersect at the point (e, e^2).

c) Find the exact value of $\displaystyle\int_1^e x^2 \ln x \,dx.$

d) Hence find, in terms of e, the area of the region shaded in the diagram.

13 The equation of a curve C is given by $y = \dfrac{x}{\cos 2x} \quad 0 < x < \dfrac{\pi}{4}.$

Find the gradient of the curve at the point where $x = \dfrac{\pi}{8}.$

14 The diagram shows the graph of
$$e^x + e^{2y} = 18.$$

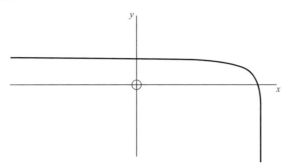

a) Find the exact co-ordinates of the points where the curve intersects the co-ordinate axes.

b) The point A has co-ordinates $(\ln 2, 2 \ln 2)$. Prove that A lies on the curve
$$e^x + e^{2y} = 18.$$

c) Find the gradient of the curve $e^x + e^{2y} = 18$ at the point A and hence show that the equation of the normal to the curve at A is $y - 16x + 14 \ln 2 = 0$.

15 If $y = x \sin 2x$ find

a) the gradient of the curve at the point where $x = \dfrac{\pi}{2}$

b) $\displaystyle\int_0^{\frac{1}{2}\pi} y \, dx.$

16 Use the substitution $u = 4 + 3 \ln x$ to evaluate $\displaystyle\int \frac{4 + 3 \ln x}{x} \, dx.$

17 Show that the exact value of $\displaystyle\int_{\frac{1}{8}\pi}^{\frac{1}{4}\pi} x \sin 4x \, dx$ is $\dfrac{1}{16}(\pi - 1).$

(OCR Mar 2000 P3)

18 i) Prove that $\cos^4 \theta - \sin^4 \theta \equiv \cos 2\theta$.

A curve is defined by the parametric equations
$$x = \cos^3 t, \ y = \sin^3 t \qquad 0 < t < \frac{1}{4}\pi.$$

ii) Prove that the normal to the curve at the point $P(\cos^3 p, \sin^3 p)$ is
$$x \cos p - y \sin p = \cos 2p.$$

The normal at P meets the x-axis at A and the y-axis at B.

iii) Prove that the line segment AB has length $2 \cot 2p$.

19 A curve C has parametric equations

$$x = \cos\theta, \ y = \cos 3\theta \qquad \text{where } 0 \leqslant \theta \leqslant \pi.$$

i) Expand $\cos(2\theta + \theta)$ and hence show that the equation satisfied by all points on C is $y = 4x^3 - 3x$.

ii)

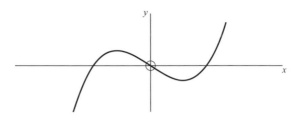

The diagram shows the graph of $y = 4x^3 - 3x$. Explain briefly why not all the points on this graph belong to C, and show on a sketch the part of the graph that corresponds to C.

iii) Find $\dfrac{dy}{dx}$ in terms of x. Use the parametric equations to find $\dfrac{dy}{dx}$ in terms of θ and hence show that

$$\sin 3\theta = \sin\theta(3 - 4\sin^2\theta).$$

(OCR Jan 2004 P3)

20 i) By differentiating $\dfrac{1}{\cos\theta}$ with respect to θ, prove that $\dfrac{d}{d\theta}(\sec\theta) = \sec\theta\tan\theta$.

ii) Use the substitution $x = \sec\theta$ to show that

$$\int_{\sqrt{2}}^{2} \sqrt{\left(1 - \frac{1}{x^2}\right)}\, dx = \int_{\frac{1}{4}\pi}^{\frac{1}{3}\pi} \tan^2\theta\, d\theta.$$

iii) Hence find the exact value of $\displaystyle\int_{\sqrt{2}}^{2} \sqrt{\left(1 - \frac{1}{x^2}\right)}\, dx.$

(OCR Jan 2004 P3)

21 Find the exact value of $\displaystyle\int_{1}^{2} x^2 \ln x \, dx.$

22 i) Show that the substitution $u = e^x + 2$ transforms

$$\int \frac{1}{e^x + 4 + 4e^{-x}}\, dx \quad \text{into} \quad \int \frac{1}{u^2}\, du.$$

ii) Hence find the exact value of $\displaystyle\int_{0}^{\ln 2} \frac{1}{e^x + 4 + 4e^{-x}}\, dx.$

(OCR Mar 2000 P3)

23 The equation of the curve shown in the diagram is

$$y = (3 - x)e^{-\frac{1}{2}x}.$$

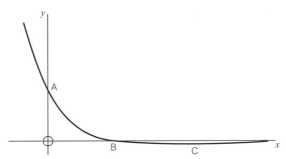

The curve cuts the y-axis at A and the x-axis at B. The point C is a stationary point.

i) Write down the co-ordinates of A and B.

ii) Calculate the co-ordinates of C.

iii) Find, in terms of e, the area bounded by the axes and the curve between the points A and B.

24 The diagram shows the curve $y = \dfrac{6 \cos 3x}{3 - \sin 3x}$.

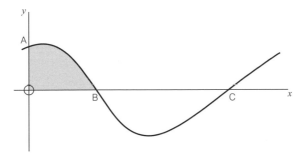

The curve cuts the y-axis at A and cuts the positive x-axis for the first time at B.

i) Write down the co-ordinates of the points A and B.

ii) Prove that the normal at B has equation $54y - 6x + \pi = 0$.

iii) Prove that the area of the shaded region is $2 \ln\left(\dfrac{3}{2}\right)$.

25 Show that $\displaystyle\int_0^{\frac{1}{4}\pi} x^2 \cos 2x \, dx = \frac{1}{32}(\pi^2 - 8)$.

(OCR Nov 1999 P3)

26 Find

i) $\displaystyle\int \ln(3x) \, dx$

ii) the exact value of $\displaystyle\int_{\frac{1}{6}\pi}^{\frac{1}{3}\pi} \sin^2 x \, dx$.

(OCR Jun 1999 P3)

27 Find

    i) $\displaystyle\int x \sin x\, dx$

    ii) the exact value of $\displaystyle\int_{\frac{1}{12}\pi}^{\frac{1}{6}\pi} \sec^2(2x)\, dx.$

(OCR Mar 1999 P3)

28 **a)** Show that

$$\int \cos^2 x\, dx = \frac{1}{2}x + \frac{1}{4}\sin 2x + c$$

    where c is a constant.

    b)

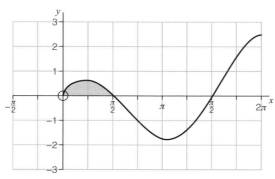

The diagram shows a sketch of the curve which has equation $y = \sqrt{x}\cos x$.
The shaded region is rotated through 2π radians about the x-axis to form a solid.

Show that the volume of this solid is $\dfrac{\pi}{16}(\pi^2 - 4)$.

29 **a)** Find the equation of the tangent to the curve $y = \tan 2x$ at the point where $x = \dfrac{1}{8}\pi$.

    b) Find the area of the closed region bounded by the curve $y = \tan 2x$, the x-axis and the line $x = \dfrac{1}{8}\pi$.

30 The parametric equations of a curve C are

    $x = a \sin \theta,\ y = 2a \cos \theta$

where a is a positive constant and $-\pi < \theta \leqslant \pi$.

    i) Show that the equation of the tangent to C at the point with parameter θ is
$2x \sin \theta + y \cos \theta = 2a$.

    ii) This tangent passes through the point $(2a, 3a)$.

        a) Show that θ satisfies an equation of the form $5 \sin(\theta + \alpha) = 2$ and state the value of $\tan \alpha$.

        b) Hence find the two possible values of θ.

(OCR Jan 2003 P3)

31 By substituting $x = e^u$, or otherwise, find $\displaystyle\int \frac{1}{x \ln x}\,dx$.

(OCR Nov 1999 P3)

32 Sketch the curve which is given by the parametric equations

$\qquad x = 3t^2,\ y = 2t^3$.

Prove that the equation of the tangent to this curve at the point $P(3p^2, p^3)$ is

$\qquad y = px - p^3$

and find the equation of the normal to the curve at P.

Find the co-ordinates of the point Q where the tangent meets the curve again.

If the line PQ is normal to the curve at Q find the possible values of p.

Revise chapter 8 before attempting this exercise.

1 i) Find $\displaystyle\int x^2 \sin x \,\mathrm{d}x$.

ii) Find $\dfrac{\mathrm{d}}{\mathrm{d}x}(\tan x - x)$, giving your answer in its simplest form.

iii) Find the general solution of the differential equation

$$\frac{\mathrm{d}y}{\mathrm{d}x} = x^2 \sin x \cot^2 y.$$

(OCR Jan 2002 P3)

2 A balloon seller uses a gas cylinder to inflate spherical balloons. A balloon has radius r at time t seconds, and it takes 4 seconds for the radius of the balloon to reach 21 cm from its initial value of 1 cm. It is desired to predict the time, T seconds, it would take to inflate the balloon to only 19 cm. In a simple model, the rate of increase of r is taken to be proportional to $\dfrac{1}{r^2}$. Express this statement as a differential equation, and find T, giving your answer correct to 3 significant figures.

(OCR Mar 1999 P3)

3 A metal rod is 60 cm long and is heated at one end. The temperature at a point on the rod at distance x cm from the heated end is denoted by T °C. At a point half way along the rod, $T = 290$ and $\dfrac{\mathrm{d}T}{\mathrm{d}x} = -6$.

i) In a simple model for the temperature of the rod, it is assumed that $\dfrac{\mathrm{d}T}{\mathrm{d}x}$ has the same value at all points on the rod. For this model, express T in terms of x and hence determine the temperature difference between the ends of the rod.

ii) In a more refined model, the rate of change of T with respect to x is taken to be proportional to x. Set up a differential equation for T, involving a constant of proportionality k.
Solve the differential equation and hence show that, in this refined model, the temperature along the rod is predicted to vary from 380 °C to 20 °C.

(OCR Jun 1999 P3)

4 The variables x and t are such that the rate of increase of $\ln x$ with respect to t is proportional to x.

i) Express this statement as an equation, and hence show that the rate of increase of x with respect to t is proportional to x^2.

ii) Find the general solution of the differential equation

$$\frac{\mathrm{d}x}{\mathrm{d}t} = kx^2$$

where k is a constant, expressing x in terms of t in your answer.

iii) Given that $x = 1$ when $t = 0$ and that $x = 2$ when $t = 1$, find the value of t near which x becomes very large.

(OCR Jun 2004 P3)

5 Paint is dropped onto absorbent paper and forms a circular mark which has radius r cm and area A cm^2 at time t seconds. When $t = 0$, $r = 0$ and when $t = 5$, $r = 2$. It is desired to predict the value of t when $A = 25$.

i) It is assumed that the rate of increase of r is constant. With this assumption, find the value of t when $A = 25$.

ii) In an alternative model, it is assumed that the rate of increase of A is inversely proportional to r. For this model, find the value of t when $A = 25$.

(OCR Mar 1998 P3)

6 In a certain pond, the rate of increase of the number of fish is proportional to the number of fish, n, present at time t. Assuming that n can be regarded as a continuous variable, write down a differential equation relating n and t, and hence show that

$$n = A\mathrm{e}^{kt},$$

where A and k are constants.

In a revised model, it is assumed also that the fish are removed from the pond, by anglers and by natural wastage, at the constant rate of p per unit time, so that

$$\frac{\mathrm{d}n}{\mathrm{d}t} = kn - p.$$

Given that $k = 2$, $p = 100$ and that initially there were 500 fish in the pond, solve this differential equation, expressing n in terms of t.

Give a reason why this revised model is not satisfactory for large values of t.

(OCR Mar 1997 P3)

7 A biologist studying fluctuations in the size of a particular population decides to investigate a model for which

$$\frac{\mathrm{d}P}{\mathrm{d}t} = kP \cos kt,$$

where P is the size of the population at time t days and k is a positive constant.

i) Given that $P = P_0$ when $t = 0$, express P in terms of k, t and P_0.

ii) Find the ratio of the maximum size of the population to its minimum size.

(OCR Jun 1996 P3)

8 Show that

$$\frac{d}{du}(\ln \tan u) = \frac{2}{\sin 2u}.$$

The variables x and y are related by the differential equation

$$\frac{dy}{dx} = \sin x \sin 2y \qquad \left(0 < y < \frac{1}{2}\pi\right)$$

and $y = \frac{1}{4}\pi$ when $x = 0$. Show that, when $x = \pi$, the value of y is $\tan^{-1}(e^4)$.

(OCR Mar 1998 P3)

9 A liquid is heated in such a way that energy is supplied to the liquid at a constant rate but energy escapes from the liquid at a rate that is proportional to the difference between the liquid's temperature and the temperature of the surroundings. The rate of change of the temperature is proportional to the rate at which the liquid gains heat.
If the heat of the surroundings is 20 °C, and the temperature at time t mins is θ °C, show that there are positive constants α and β such that

$$\frac{d\theta}{dt} = \alpha - \beta(\theta - 20).$$

When $t = 0$, $\theta = 20$, $\frac{d\theta}{dt} = 15$ and when $\theta = 35$, $\frac{d\theta}{dt} = 12$.

i) Determine the values of α and β.

ii) Solve the differential equation and find the time when $\theta = 35$.

iii) Describe what happens to the temperature of the liquid as the heating continues.

10 a) If $V = \pi r^2 h$ where h is a constant, show that $\frac{dV}{dt} = 2\pi rh \frac{dr}{dt}$.

In a leakage from an oil rig, oil falls continuously onto a smooth sea at a constant rate of 0.04π m³/minute. It spreads into a thin circular slick of constant depth 2 cm and at time t minutes after the leakage started the radius of the film is r metres.
Simultaneously, the oil evaporates at a rate of $10^{-6}A$ m³/min where A is the surface area of the slick.

b) Explain why the volume, V, of oil in the slick, the radius, r, of the slick and the time t are related by the differential equation

$$\frac{dV}{dt} = 0.04\pi - 10^{-6}\pi r^2.$$

c) Deduce that r is related to t by the differential equation

$$40\,000r \frac{dr}{dt} = 40\,000 - r^2 \qquad t = 0, r = 0.$$

d) Solve this differential equation and show that no matter how long the leak continues, the slick will never exceed a radius of 200 m.

e) How many days must the rig be leaking for the radius of the slick to exceed 150 m?

11 A rectangular water tank has a horizontal square base of side 1 m. Water is flowing out of the tank from an outlet in the base but is also being pumped into the tank at a constant rate of 400 cm^3 s^{-1}. Initially the depth of the water in the tank is 81 cm and water is flowing out at a rate of 900 cm^3 s^{-1}.

i) Assuming that the rate at which the water is flowing out remains constant, calculate the time, in seconds, taken for the depth of the water to decrease from 81 cm to 64 cm.

ii) In an improved model, it is assumed that the water flows from the outlet at a rate which is proportional to the square root of the depth, h cm, of water in the tank.

    a) Show how this model leads to the differential equation

$$\frac{dh}{dt} = 0.04 - 0.01\sqrt{h}$$

    where t is time in seconds.

    b) Show that the solution of the differential equation in (a) is given by

$$t = \int \frac{100}{4 - \sqrt{h}}\, dh.$$

    Use the substitution $u = \sqrt{h} - 4$ to find the time, in seconds, for the depth of the water to decrease from 81 cm to 64 cm.

(OCR Nov 1998 P3)

12 The by-products of an explosive reaction are two chemicals, X and Y. When x grams of X have been produced, y grams of Y have been produced. The rate of change of x with respect to time is proportional to e^x and the rate of change of y with respect to time is proportional to e^y. Show that

$$\frac{dy}{dx} = ke^{y-x},$$

for some constant k.

Initially no X or Y is present and at a later instant the amounts, in grams, of X and Y that have been produced are ln 5 and ln 10 respectively. Solve the above differential equation to find a formula for y in terms of x.

By how much does the amount of Y exceed the amount of X at an instant when the amount of Y is increasing three times as rapidly as the amount of X?

(OCR Jun 2000 P3)

Revise chapter 4 before attempting this exercise.

1 The diagram shows two vectors, **a** and **b**. The vector **a** is 7 units long and the vector **b** is 4 units long and the angle between the two vectors is 60°.

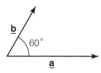

a) Draw a diagram showing the vectors **a**, **b**, **a** + 2**b** and 3**a** − 2**b**.

b) Calculate the length of the vector 3**a** − 2**b**.

c) Write down the value of **a** · **b**.

2 **a)** Write down the vector equation of the line, L, joining the points A(6, 3, 0) and B(8, 7, 4).

The line M has equation $\mathbf{r} = \begin{pmatrix} -1 \\ -3 \\ -8 \end{pmatrix} + s\begin{pmatrix} 3 \\ 2 \\ 3 \end{pmatrix}$.

b) Prove that the lines L and M intersect and find the co-ordinates of the point of intersection.

c) By using an appropriate scalar product, or otherwise, find the acute angle between the two lines, giving your answer to the nearest tenth of a degree.

3 The position vectors of the points A and B, relative to a fixed origin O, are 6**i** + 4**j** − **k** and 8**i** + 5**j** − 3**k** respectively.

i) Find $\overrightarrow{\mathbf{AB}}$.

ii) Find the length AB.

iii) Show that, for all values of the parameter λ, the point P with position vector

$(8 + 2\lambda)\mathbf{i} + (5 + \lambda)\mathbf{j} - (3 + 2\lambda)\mathbf{k}$

lies on the straight line through A and B.

iv) Determine the value of λ for which $\overrightarrow{\mathbf{OP}}$ is perpendicular to $\overrightarrow{\mathbf{AB}}$.

(OCR Jan 2002 P3)

4 The floor of a room is a rectangle OABC in which OA = 16 m and OC = 8 m. The ceiling is 4 m above the floor and its corner points, K, L, M, N, are vertically above O, A, B, C respectively.

Axes are chosen such that OA, OC and OD are along Ox, Oy and Oz respectively.

i) Write down the vector equation of the line CL.

If P is the midpoint of KN

ii) write down the co-ordinates of P;

iii) write down the vector equation of the line BP;

iv) prove that the lines CL and BP are skew.

v) Calculate the value of the scalar product $\overrightarrow{OM} \cdot \overrightarrow{OP}$.

vi) Determine, correct to one decimal place, the angle MOP.

5 Two lines have equations

$$\mathbf{r} = \begin{pmatrix} 3 \\ 1 \\ -2 \end{pmatrix} + s \begin{pmatrix} 1 \\ -1 \\ 4 \end{pmatrix} \quad \text{and} \quad \mathbf{r} = \begin{pmatrix} 1 \\ 0 \\ 5 \end{pmatrix} + t \begin{pmatrix} -2 \\ -3 \\ 1 \end{pmatrix}.$$

i) Find the acute angle between the directions of the two lines.

ii) Prove that the lines are skew.

(OCR Jan 2003 P3)

6 The equations of the lines L_1 and L_2 are

$$\mathbf{r} = \begin{pmatrix} 9 \\ 2 \\ -1 \end{pmatrix} + s \begin{pmatrix} 3 \\ -4 \\ 0 \end{pmatrix} \quad \text{and} \quad \mathbf{r} = \begin{pmatrix} 2 \\ 3 \\ 2 \end{pmatrix} + t \begin{pmatrix} a \\ 3 \\ b \end{pmatrix}.$$

It is given that L_1 and L_2 are perpendicular and also that L_1 and L_2 intersect.

i) Find the values of a and b.

ii) Hence show that the point of intersection A has position vector $\begin{pmatrix} 6 \\ 6 \\ -1 \end{pmatrix}$.

iii) There are two points on L_1 whose distance from A is 10. Find the position vector of one of these points.

(OCR Jun 2003 P3)

7 The position vectors of A and B are $\begin{pmatrix} 4 \\ -3 \\ 2 \end{pmatrix}$ and $\begin{pmatrix} 2 \\ -2 \\ -7 \end{pmatrix}$ respectively.

    i) Prove that the lines OA and OB are perpendicular.

    ii) Find the vector $\overrightarrow{\mathbf{AB}}$ and deduce the length of AB.

    iii) Write down the vector equation of the line AB. Prove that the line AB intersects the line whose vector equation is

$$\mathbf{r} = \begin{pmatrix} 3 \\ -1 \\ 2 \end{pmatrix} + t \begin{pmatrix} -1 \\ 1 \\ -3 \end{pmatrix}$$

    and find the co-ordinates of the point of intersection.

8 The lines L_1 and L_2 have vector equations

$$\mathbf{r} = \begin{pmatrix} 1 \\ 4 \\ -3 \end{pmatrix} + s \begin{pmatrix} 2 \\ -2 \\ 1 \end{pmatrix} \quad \text{and} \quad \mathbf{r} = \begin{pmatrix} 8 \\ 2 \\ 5 \end{pmatrix} + t \begin{pmatrix} 3 \\ 2 \\ 6 \end{pmatrix}$$

    respectively, where s and t are parameters.

    i) Find the acute angle between the directions of the two lines.

    ii) Prove that the lines intersect and find the position vector of their point of intersection.

9 The lines L_1 and L_2 have vector equations

$$\mathbf{r} = \begin{pmatrix} 3 \\ 2 \\ 4 \end{pmatrix} + s \begin{pmatrix} 1 \\ 3 \\ -5 \end{pmatrix} \quad \text{and} \quad \mathbf{r} = \begin{pmatrix} -3 \\ 4 \\ 6 \end{pmatrix} + t \begin{pmatrix} 1 \\ -2 \\ 2 \end{pmatrix}$$

    respectively, where s and t are parameters.

    i) Find the acute angle between the directions of the two lines.

    ii) Prove that the lines intersect and find the position vector of their point of intersection.

1 The points A and B have position vectors $4\underline{i} - 2\underline{j} + 3\underline{k}$ and $\underline{i} + \underline{j} - 3\underline{k}$ respectively. Find a unit vector parallel to the line segment $\overset{\rightarrow}{AB}$. [4]

2 The equation of a curve is given by

$$\sin y - x = \cos 2x, \qquad \text{where} \quad \frac{\pi}{4} < x < \frac{\pi}{2} \quad \text{and} \quad 0 < y < \frac{\pi}{2}.$$

Use implicit differentiation to find $\dfrac{dy}{dx}$ in terms of x and y, and hence find the exact value of the x co-ordinate of the stationary point on the curve. [6]

3 i) Simplify $f(x) = \dfrac{12 + x - x^2}{6 - x - x^2}$. [2]

ii) Hence expand $f(x)$ in ascending powers of x, up to and including the term in x^2. [5]

4 Use the substitution $u = e^x + 1$ to find the exact value of $\displaystyle\int_0^1 \frac{e^x(e^x - 2)}{e^x + 1}\,dx$. [7]

5 i) Find the quotient and remainder when $x^4 + 5x^3 + 5x^2 - x - 7$ is divided by $(x + 1)^2$. [4]

ii) Hence, or otherwise, differentiate $g(x) = \dfrac{x^4 + 5x^3 + 5x^2 - x - 7}{(x + 1)^2}$. [3]

6 Find

a) $\displaystyle\int (1 + \tan x)^2\,dx$ [4]

b) $\displaystyle\int x^4 \ln x\,dx$ [4]

7 Two coplanar lines have the vector equations

$$\underline{r} = 2\underline{i} + 3\underline{j} + \underline{k} + \lambda(\underline{i} + 3\underline{j} + 2\underline{k})$$

$$\text{and} \quad \underline{r} = 7\underline{i} + 3\underline{j} + 5\underline{k} + \mu(-\underline{i} + 2\underline{j}).$$

i) Find the point of intersection of the lines. [5]

ii) Find the acute angle between the lines. [5]

8 In a certain chemical reaction the differential equation relating the amount, x, of one of the substances to the time, t, is

$$\frac{dx}{dt} = k(a - x)(2a - x),$$

where k and a are constants and where $0 \leqslant x \leqslant a$.

It is given that $x = 0$ when $t = 0$ and that $x = \dfrac{2}{3}a$ when $t = 1$.

Solve the equation and show that $k = \dfrac{1}{a}\ln 2$. [10]

9 A curve is given parametrically by the equations

$$x = \frac{1}{2}\left(t + \frac{1}{t}\right), \qquad y = \frac{1}{2}\left(t - \frac{1}{t}\right).$$

i) Show that the gradient of the curve at the point P, with parameter $t = 3$, has the value 5/4. [3]

ii) The normal to the curve at P intersects the curve again at Q. Find the co-ordinates of Q. [7]

iii) Find the Cartesian equation of the curve. [3]

CHAPTER 1
Exercise 1

1
 a) $-4 \sin x$
 b) $5 - 14 \sin 2x$
 c) $5 \cos x - 3 \sin x$
 d) $-32 \sin 4x - 12 \cos 4x$
 e) $3 + 35 \sin 5x$

2 $\dfrac{dx}{dt} = -3\sqrt{3}$

3 $\dfrac{dy}{dx} = 4 - 3\sqrt{2}$

5
 a) $A\left(\dfrac{\pi}{2}, \dfrac{\pi}{2}\right)$ $B(\pi, \pi)$

 b) $\dfrac{dy}{du} = 1 + 2 \cos 2x$

 c) $C\left(\dfrac{\pi}{3}, \dfrac{\pi}{3} + \dfrac{\sqrt{3}}{2}\right)$ $D\left(\dfrac{2\pi}{3}, \dfrac{2\pi}{3} - \dfrac{\sqrt{3}}{2}\right)$

6 $x = \dfrac{\pi}{6}$ max; $x = \dfrac{\pi}{3}$ min;

 $x = \dfrac{7\pi}{6}$ max; $x = \dfrac{4\pi}{3}$ min

CHAPTER 1
Exercise 2

1
 a) $5x^4 \cos(x^5)$
 b) $4x^3 \sin 2x + 2x^4 \cos 2x$
 c) $\dfrac{5 \cos 5x(1 + x^2) - 2x \sin 5x}{(1 + x^2)^2}$
 d) $2e^{2x} \cos 3x - 3e^{2x} \sin 3x$
 e) $5 \cos x \sin^4 x$
 f) $-6 \sin 2x \cos^2 2x$
 g) $\dfrac{1}{1 + \cos x}$
 h) $10 \sin 5x(3 + \cos 5x)^{-3}$

2 -6

3 $k = \dfrac{1}{4}$

5 2π

7
 a) $-7 \cot x \operatorname{cosec} x - 3 \operatorname{cosec}^2 x$
 b) $10 \sec 2x \tan 2x + 2 \sec^2 2x$
 c) $3x^2 \tan 3x + 3x^3 \sec^2 3x$
 d) $5 \tan x \sec^5 x$
 e) $2e^{2x} \sec 5x + 5e^{2x} \sec 5x \tan 5x$
 f) $\dfrac{2 \sec 2x(\tan 2x - 1)}{(1 + \tan 2x)^2}$

8 **i)** $A\left(\dfrac{\pi}{6}, 0\right)$ **iv)** $\theta = 0.287$

9 $\dfrac{dy^2}{dt^2} = 12e^{-3t} \sin 3t - 5e^{-2t} \cos 3t$

10 $f'(0) = 1$
 $f''(0) = 0$

CHAPTER 1
Exercise 3

1
 a) $-\dfrac{1}{3} \cos 3x + c$
 b) $-\dfrac{4}{3} \cos 3x + 4 \sin 2x + c$
 c) $-\dfrac{1}{5} \cos 5x - \dfrac{5}{3} \sin 3x + c$
 d) $2 \sin 2x - 2 \sin 4x + c$

2 $x = 5t + 3 \sin t + 2 \quad \dfrac{5\pi}{2} + 5$

3 **a)** 1 **b)** $\dfrac{1}{2}\pi^2 + 2$ **c)** $-\dfrac{1}{3}$

4 $\dfrac{\sqrt{3}}{8}$

5 2

6 $\dfrac{1}{4}\pi(4 - \pi)$

7
 a) $y = 4 \sin x - 3 \cos x + c$
 b) $y = 3 \sin 3x + 4 \cos 2x + c$

8 $4 \sin 3x - 2 \cos 4x + c$

9 $\sqrt{3}\pi$

10 **a)** 0 **b)** 2

11 $2\sqrt{3} - \dfrac{2\pi}{3}$

12 $\left(\dfrac{2\pi}{3}, \dfrac{3}{2}\right) \left(\dfrac{4\pi}{3}, \dfrac{3}{2}\right)$

 Area $= 2\sqrt{3} - \dfrac{2\pi}{3}$

CHAPTER 1
Exercise 4

1 **a)** $\frac{1}{2}x + \frac{1}{4}\sin 2x + c$

   b) $\frac{1}{2}x - \frac{1}{4}\sin 2x + c$

   c) $\frac{1}{2}x + \frac{1}{16}\sin 8x + c$

   d) $\frac{1}{2}x - \frac{1}{28}\sin 14x + c$

   e) $\frac{1}{2}x - \frac{1}{20}\sin 10x + c$

2 **a)** $\frac{\pi}{16} - \frac{1}{8}$ **b)** $\frac{\pi}{24} + \frac{1}{12}$

3 $\frac{3}{2}\pi - 4$

4 $\frac{9\pi^2}{2}$

5 **c)** $\frac{3\pi}{16}$ **d)** $\frac{3\pi}{16}$

CHAPTER 1
Revision Exercise

1 $\frac{dy}{dx} = -10\sin 2x + 6\cos 2x$

   $\frac{dy^2}{dx^2} = -20\cos 2x - 12\sin 2x$

2 $\frac{x\sec^2 x - \tan x}{x^2}$

3 **a)** $2x\cos 4x - 4x^2\sin 4x$

   b) $\frac{5\sqrt{2}}{6} + 2$

4 $\frac{5\sqrt{3}}{2}$

5 **i)** 1

6 **a)** $2 + \sqrt{3}$ **b)** $\pi\left[\frac{10\pi}{3} + \sqrt{3}\right]$

7 -30

8 $\left(\frac{\pi}{4}, \frac{1}{\sqrt{2}}\,e^{-\pi/4}\right)$ $\left(\frac{5\pi}{4}, \frac{1}{\sqrt{2}}\,e^{-5\pi/4}\right)$

9 $y = 4\sqrt{3}x + 2 - \frac{2\pi\sqrt{3}}{3}$

10 $(0, 5)$

11 **a)** $f'(x) = -6\sin 6x$

    b) $g'(x) = 12\sin 3x\cos 3x$

12 **b)** $y = x - \frac{3\pi}{2}$

13 **a)** $2t\tan 3t + 3t^2\sec^2 3t$

    b) $\frac{-6\sin 3t}{2 + \cos 3t}$

14 $\frac{dy}{dx} - \frac{1}{x^2}\sec x + \frac{1}{x}\sec x\tan x$

    $\theta = 0.860$

CHAPTER 2
Exercise 1

1 $\frac{4x^3}{x^4 + 1}$ **2** $\frac{5}{5x - 3}$

3 $\frac{2x}{x^2 + 4}$ **4** $\frac{3\cos x}{5 + 3\sin x}$

5 $\frac{-4\sin 4x}{2 + \cos 4x}$ **6** $\frac{3e^{3x}}{e^{3x} - 5}$

CHAPTER 2
Exercise 2

1 $\ln(x^2 + 4) + c$

2 $\ln|4x - 3| + c$

3 $\ln|x^3 + 16| + c$

4 $\ln|3 - \cos 2x| + c$

5 $6\ln(1 + e^x) + c$

6 $3\ln(5 + 2\sin x) + c$

7 $-5\ln(3 + \cos 2x) + c$

8 $2\ln(3e^{2x} + 4) + c$

9 $\frac{1}{2}\ln(x^2 + 16)$

10 $\frac{1}{2}\ln(x^2 + 2x + 4) + c$

11 $\frac{1}{5}\ln|\sin 5x| + c$

12 $\frac{1}{3}\ln(1 + e^{3x}) + c$

13 $-\frac{1}{3}\ln(4 + e^{-3x}) + c$

14 $\frac{1}{2}\ln 2$ **15** $\frac{1}{2}\ln 2$

16 $\frac{1}{2}\ln\left(\frac{5}{3}\right)$ **17** $3\ln\left(\frac{1}{2}\right)$

18 $\frac{1}{2}\ln\left(\frac{5}{12}\right)$ **19** $\frac{1}{2}\ln 5$

20 $3 + \dfrac{7}{x-3}$ $6 + 7\ln 3$

21 a) $(0, -2)$ $(4, 6)$
    b) Area $= \dfrac{15}{2} - 4\ln 4$

22 $\dfrac{7x+2}{x(x+1)}$ $5\ln 2 + 2\ln 3$

23 $\dfrac{2x-13}{(2x+1)(x-3)}$
    $2\ln|2x+1| - \ln|x-3| + c$

CHAPTER 2
Exercise 3

1 a) $\dfrac{4}{x+5} - \dfrac{1}{x+3}$ **b)** $\dfrac{2}{x+3} + \dfrac{2}{x-3}$

    c) $\dfrac{2}{x-1} + \dfrac{1}{3x+5}$

2 $3\ln|x+2| - 2\ln|x-3| + c$

3 $\ln\left(\frac{3}{2}\right)$

4 a) $\dfrac{9}{x-1} - \dfrac{1}{x-3}$

    b) $\dfrac{-9}{(x-1)^2} + \dfrac{1}{(x-3)^2}$

    c) $(4, 2)$ $(2.5, 8)$
    d) Area $= 9\ln 3 - 10\ln 2$

5 a) $x = -1, \ x = 2, \ y = 0.$
    b) Area $= 5\ln 2$

CHAPTER 2
Exercise 4

1 $\frac{1}{14}(2x-5)^7 + c$

2 $\frac{1}{16}(4x+1)^4 + c$

3 $\frac{1}{9}(6x+5)^{\frac{3}{2}} + c$

4 $\frac{1}{30}(2x^3+1)^5 + c$

5 $\frac{1}{6}(1+\cos x)^6 + c$

6 $-\frac{1}{8}(3+\cos 2x)^4 + c$

7 $\frac{1}{3}(4+e^{2x})^{\frac{3}{2}} + c$

8 $\dfrac{1}{16}$ **9** $\dfrac{10}{3}$ **10** 1

11 $\dfrac{1}{10}$ **12** $\dfrac{7}{54}$ **13** $\dfrac{88}{3}$

14 10 **15** $\frac{1}{2}\ln\left(\frac{25}{16}\right) - \frac{9}{50}$

CHAPTER 2
Revision Exercise

1 i) $\left(2, \frac{1}{4}\right)$ max $\left(-2, -\frac{1}{4}\right)$ min
    ii) Area $= \dfrac{1}{2}\ln 2$

2 1

3 $\dfrac{8}{1+4x} + \dfrac{3}{1-3x}$

4 $\frac{1}{4}\ln 5$

5 $\frac{1}{18}(1+3e^{2x})^3 + c$

6 $\ln 2 - \dfrac{1}{2}$

8 a) $\frac{1}{4}\ln\left(\frac{5}{3}\right)$ **b)** $\frac{1}{2}(\sqrt{5} - \sqrt{3})$

10 a) $\frac{1}{2}\ln 2$ **b)** $\dfrac{1}{36}$

CHAPTER 3
Exercise 1

1 i) $1 - 2x + 3x^2 - 4x^3 + 5x^4 - \cdots$

2 ii) $1 - \frac{1}{4}x + \frac{5}{32}x^2 - \frac{15}{128}x^3 + \frac{195}{2048}x^4 + \cdots$

3 i) $1 + \frac{3}{2}x + \frac{3}{8}x^2 - \frac{x^3}{16} + \frac{3x^4}{128} \cdots$

CHAPTER 3
Exercise 2

1 $1 - \frac{1}{2}x + \frac{3}{8}x^2 - \frac{5}{16}x^3 + \cdots$ $-1 < x < 1$

2 $1 - 3x + 6x^2 - 10x^3 + \cdots$ $-1 < x < 1$

3 $1 + 4x + 16x^2 + 64x^3 + \cdots$ $-\frac{1}{4} < x < \frac{1}{4}$

4 $\frac{1}{4} - \frac{1}{4}x + \frac{3}{16}x^2 - \frac{1}{8}x^3 + \cdots$ $-2 < x < 2$

5 $3 + \frac{1}{6}x - \frac{1}{216}x^2 + \frac{1}{3888}x^3 + \cdots$ $-9 < x < 9$

6 $\frac{1}{4} + \frac{3}{4}x + \frac{27}{16}x^2 + \frac{27}{8}x^3 + \cdots$ $\quad -\frac{2}{3} < x < \frac{2}{3}$

7 $(1-2x)^{-2} = 1 + 4x + 12x^2$
$$+ 32x^3 + 80x^4 + \cdots$$
$(2+3x)(1-2x)^{-2} = 2 + 11x + 36x^2$
$$+ 100x^3 + 256x^4 + \cdots$$
$$-\frac{1}{2} < x < \frac{1}{2}$$

8 $2 + 3x + 7x^2 + \frac{69}{4}x^3 + \cdots$

9 $5 + \frac{1}{10}x - \frac{1}{1000}x^2 + \cdots$ $\quad -25 < x < 25$

10 a) $\dfrac{1}{1-x} + \dfrac{3}{1+2x}$
b) i) $1 + x + x^2 + x^3 + \cdots$
ii) $1 - 2x + 4x^2 - 8x^3 + \cdots$
c) $4 - 5x + 13x^2 - 23x^3 + \cdots$

11 Both are $1 + 2x - 6x^2$
$p = 55, q = 53$

CHAPTER 3
Revision Exercise

1 $1 - 8x + 48x^2 - 256x^3 + \cdots$ $\quad -\frac{1}{4} < x < \frac{1}{4}$

2 $5 + \frac{1}{10}x - \frac{1}{1000}x^2 + \cdots$ $\quad -25 < x < 25$
$\sqrt{2510} \approx 50.0999$

3 $\dfrac{5}{1-x} + \dfrac{3}{1+3x}$
$8 - 4x + 32x^2 - 76x^3 + \cdots$ $\quad -\frac{1}{3} < x < \frac{1}{3}$

4 $1 + 2x - 6x^2 + 28x^3 + \cdots$ $\quad p = -5$

5 $q = -2$
$1 + 8x + 40x^2 + 160x^3 + \cdots$ $\quad -\frac{1}{2} < x < \frac{1}{2}$

6 a) i) $1 + \frac{1}{2}px - \frac{1}{8}p^2x^2 + \frac{1}{16}p^3x^3 - \frac{5}{128}p^4x^4$
ii) $1 - \frac{1}{2}qx^2 + \frac{3}{8}q^2x^4$
b) $p = 2 \qquad q = -4$
$\frac{5}{2}x^3 + \frac{19}{8}x^4 + \cdots$

7 i) $3 + \frac{1}{27}x - \frac{1}{2187}x^2$ $\quad -27 < x < 27$

8 $a = \pm 12 \qquad p = 5, a = -2$

CHAPTER 4
Exercise 1

3 a) $\begin{pmatrix} 11 \\ -10 \end{pmatrix}$ **b)** $\begin{pmatrix} 5 \\ 11 \end{pmatrix}$ **c)** $\begin{pmatrix} 3 \\ 3 \end{pmatrix}$

5 20.45 $\qquad$ 22.2°

6 12.4°

7 $\overrightarrow{OE} \quad \frac{1}{2}\mathbf{a} + \frac{1}{2}\mathbf{c}$
$\overrightarrow{OF} \quad \frac{3}{2}\mathbf{a} + \frac{1}{2}\mathbf{c}$
OEFA is parallelogram

CHAPTER 4
Exercise 2

1 Possible answers
a) $\mathbf{r} = \begin{pmatrix} 0 \\ 3 \end{pmatrix} + t\begin{pmatrix} 1 \\ 5 \end{pmatrix}$ **b)** $\mathbf{r} = \begin{pmatrix} 0 \\ 2 \end{pmatrix} + t\begin{pmatrix} 1 \\ -3 \end{pmatrix}$
c) $\mathbf{r} = \begin{pmatrix} 3 \\ 0 \end{pmatrix} + t\begin{pmatrix} 0 \\ 1 \end{pmatrix}$ **d)** $\mathbf{r} = \begin{pmatrix} 2 \\ 7 \end{pmatrix} + t\begin{pmatrix} 2 \\ 3 \end{pmatrix}$
e) $\mathbf{r} = \begin{pmatrix} -1 \\ 4 \end{pmatrix} + t\begin{pmatrix} 4 \\ 1 \end{pmatrix}$

2 a) $y = 5x - 11$ **b)** $y = -\frac{1}{2}x + \frac{13}{2}$

3 a) $(5, 6)$ **b)** $\left(-\frac{3}{5}, \frac{2}{5}\right)$
c) Two lines are the same
d) Lines are parallel and distinct
e) $(7, 1)$ **f)** $(3, 9)$

CHAPTER 4
Exercise 3

1 a) $\begin{pmatrix} 4 \\ 10 \\ -10 \end{pmatrix}$ **b)** $\begin{pmatrix} 14 \\ 1 \\ 10 \end{pmatrix}$
c) $(11\mathbf{i} + \mathbf{j} + 2\mathbf{k})$

2 a) 13 **b)** $\sqrt{17}$ **c)** $\sqrt{38}$

3 a) $\mathbf{r} = \begin{pmatrix} 0 \\ 0 \\ 2 \end{pmatrix} + t\begin{pmatrix} 2 \\ 1 \\ -2 \end{pmatrix}$
b) $\mathbf{r} = \begin{pmatrix} 1 \\ 5 \\ 7 \end{pmatrix} + t\begin{pmatrix} 2 \\ 0 \\ 0 \end{pmatrix}$
c) $\mathbf{r} = \begin{pmatrix} 1 \\ 2 \\ 4 \end{pmatrix} + t\begin{pmatrix} 1 \\ 2 \\ 4 \end{pmatrix}$

4 a) $\dfrac{x-1}{2} = \dfrac{y-2}{1} = \dfrac{z-6}{3}$

b) $\dfrac{x-3}{-4} = \dfrac{y}{2} = \dfrac{z-6}{5}$

5 a) parallel
b) intersect at $(1, 2, 0)$
c) skew

6 $\lambda = -6$ $\quad(-4, 3, 4)$

CHAPTER 4
Exercise 4

1 a) $6\sqrt{2}$ **b)** 0 **c)** -6
d) 9 **e)** $6\sqrt{2}$ **f)** $-4\sqrt{2}$
g) 0 **h)** 4

2 a) 2 **b)** 3 **c)** -18
d) 11 **e)** 0 **f)** -12

3 0 perpendicular

4 i) $p = -2$
ii) $u = 1, v = 16$

CHAPTER 4
Exercise 5

1 i) a) $5, \sqrt{2}$ **b)** -1 **c)** $98.1°$
ii) a) $\sqrt{113}, \sqrt{452}$ **b)** 0 **c)** $90°$
iii) a) $\sqrt{14}, 3$ **b)** 6 **c)** $57.7°$

2 a) $90°$ **b)** $\theta = 15.3°$

3 a) 236.2 km/hr
b) $36, \quad 55.3°$

c) $\mathbf{r} = \begin{pmatrix} 4 \\ 2 \\ 5 \end{pmatrix} + s\begin{pmatrix} -6 \\ 5 \\ 1 \end{pmatrix}$

d) $(34, -23, 0)$

4 a) $\mathbf{r_1} = \begin{pmatrix} 1 \\ 5 \\ 3 \end{pmatrix} + s\begin{pmatrix} -1 \\ 2 \\ 1 \end{pmatrix}$

b) $\lambda = 8$ $C(-2, 11, 6)$

c) $78.6°$ **d)** $\dfrac{21\sqrt{34}}{17}$

5 a) $AB = \sqrt{18}$ $AC = \sqrt{5}$
b) -7 **c)** $\dfrac{-7}{\sqrt{90}}$ **d)** $\dfrac{\sqrt{41}}{\sqrt{90}}$

6 $\mathbf{r} = \begin{pmatrix} 4 \\ 2 \\ 6 \end{pmatrix} + t\begin{pmatrix} 3 \\ 6 \\ 3 \end{pmatrix}$ $\quad N(5, 4, 7)$ $\quad \mathbf{r} = \begin{pmatrix} 4 \\ 2 \\ 6 \end{pmatrix} + \lambda\begin{pmatrix} 5 \\ -2 \\ 5 \end{pmatrix}$

7 $p = 6$ $\theta = 90°$ 5

8 a) $30°$ **d)** $\overrightarrow{PQ} = \begin{pmatrix} 2 + q - 2p \\ -3 + p \\ 5 + q - p \end{pmatrix}$

$q = -2, p = 1$

9 $\overrightarrow{OB} \cdot \overrightarrow{AC} = 0$
diagonals of rhombus are perpendicular

CHAPTER 4
Revision Exercise

1 i) $\begin{pmatrix} 0 \\ 11 \\ -20 \end{pmatrix}$ **ii)** $\sqrt{521}$

iii) -5 **iv)** $93.4°$

2 a) $\mathbf{r} = \begin{pmatrix} 6 \\ 3 \\ 0 \end{pmatrix} + t\begin{pmatrix} 2 \\ 4 \\ 4 \end{pmatrix}$

b) $(5, 1, -2)$ **c)** 22.5

3 i) $\mathbf{r_1} = \begin{pmatrix} 3 \\ 6 \\ 1 \end{pmatrix} + s\begin{pmatrix} 2 \\ 3 \\ -1 \end{pmatrix}$

$\mathbf{r_2} = \begin{pmatrix} 3 \\ -1 \\ 4 \end{pmatrix} + t\begin{pmatrix} 1 \\ -2 \\ 1 \end{pmatrix}$

ii) $(1, 3, 2)$ **iii)** $56.9°$

4 i) -7.5 **ii)** $l = \sqrt{351}$

5 i) $T(2, -8, 20)$

ii) $\begin{pmatrix} -6 \\ -2 \\ 20 \end{pmatrix}$ $GT = \sqrt{440}$

iii) 400 **iv)** $36.3°$

6 i) $\begin{pmatrix} -3 \\ 3 \\ 1 \end{pmatrix}$ $AB = \sqrt{19}$

ii) $\mathbf{r_1} = \begin{pmatrix} 3 \\ -1 \\ 2 \end{pmatrix} + t\begin{pmatrix} -3 \\ 3 \\ 1 \end{pmatrix}$

7 ii) $77.7°$ iii) $\begin{pmatrix} 2 \\ 5 \\ 2 \end{pmatrix}$

8 ii) $p = 3$ iii) $2\sqrt{2}$

9 i) $\overrightarrow{OC} = \begin{pmatrix} 9 \\ 3 \\ 3 \end{pmatrix}$ $\overrightarrow{OD} = \begin{pmatrix} 4 \\ 8 \\ 6 \end{pmatrix}$

   ii) $\underline{r}_2 = \begin{pmatrix} 2 \\ 4 \\ 3 \end{pmatrix} + \mu \begin{pmatrix} 7 \\ 1 \\ 0 \end{pmatrix}$

   iii) $X(\frac{11}{2}, \frac{9}{2}, 3)$ iv) $81°$

10 i) $\underline{r}_1 = \begin{pmatrix} 0 \\ 0 \\ -2 \end{pmatrix} + s \begin{pmatrix} 0 \\ 1 \\ 0 \end{pmatrix}$

   ii) $B(0, 8, -2)$ iv) $29°$

CHAPTER 5
Exercise 1

1 a) $\dfrac{9}{x + 3}$ b) $\dfrac{x + 3}{x^2}$ c) $\dfrac{5x}{x^2 + 4}$

2 a) $x + 1 + \dfrac{3}{x + 2}$

   b) $2x - 12 + \dfrac{86x + 12}{x^2 + 6x}$

   c) $x^2 + 2x + 4 + \dfrac{8}{x - 2}$

   d) $3x^2 + x - 12 + \dfrac{43 - 4x}{x^2 + 4}$

   e) $x^2 - 2x + \dfrac{8x + 1}{x^2 + 2x + 4}$

3 a) $\dfrac{1}{2}x - \dfrac{3}{2} - \dfrac{3}{2x - 2}$

   b) $x - 4 + \dfrac{16}{x + 4}$

4 a) $\dfrac{9x - 3}{(x + 1)(x - 2)}$

   b) $\dfrac{2x - 1}{(x + 2)(x + 3)}$

   c) $\dfrac{5x^2 + 12x + 1}{(x + 1)(x - 2)(x + 3)}$

   d) $\dfrac{3x^2 + 14x + 19}{(x + 1)(x + 2)(x + 5)}$

   e) $\dfrac{7x^2 - x - 3}{(x + 2)(x - 1)^2}$

   f) $\dfrac{5x^2 + 4x - 117}{(2x - 2)(x + 5)^2}$

CHAPTER 5
Exercise 2

1 a) $\dfrac{2}{x + 4} + \dfrac{5}{x - 3}$ b) $\dfrac{5}{x - 2} - \dfrac{4}{x + 3}$

   c) $\dfrac{1}{2}\left[\dfrac{1}{x - 2} - \dfrac{1}{x + 4} \right]$

2 $5 \ln 3 - 3 \ln 5$

3 a) $\dfrac{5}{x + 1} + \dfrac{8}{2 - x}$

   b) $-1 < x < 1$

4 a) $\dfrac{-4}{x - 3} + \dfrac{3}{x - 2} + \dfrac{2}{x + 1}$

   b) $\dfrac{-2}{3x + 1} + \dfrac{3}{2x - 1} + \dfrac{5}{x + 1}$

5 $\dfrac{4}{2x + 1} - \dfrac{2}{x + 2} + \dfrac{1}{x + 1}$

CHAPTER 5
Exercise 3

1 a) $\dfrac{1}{x + 3} - \dfrac{1}{(x + 3)^2} + \dfrac{4}{x - 2}$

   b) $\dfrac{2}{2x + 1} - \dfrac{1}{x - 1} - \dfrac{3}{(2x + 1)^2}$

   c) $\dfrac{3}{x + 1} - \dfrac{2}{x^2} + \dfrac{1}{x}$

   d) $\dfrac{3}{3 + x} - \dfrac{2}{(3 + x)^2} + \dfrac{4}{x - 5}$

   e) $\dfrac{4}{x^2} - \dfrac{5}{x} + \dfrac{3}{2 - x}$

2 a) $\ln|1 + x| + c$ $\ln|2 + x| + c$

   b) $\dfrac{-1}{2 + x} + c$

   c) $\dfrac{1}{(2 + x)^2} + \dfrac{2}{2 + x} + \dfrac{1}{x + 1}$

   d) $2 \ln 3 - \ln 2 + \dfrac{1}{6}$

3 $\dfrac{-5}{x-2} + 3\ln|x+1| - \ln|x-2| + c$

4 a) i) $1 - x + x^2 - x^3 + \cdots$
  ii) $1 - 2x + 3x^2 - 4x^3 + \cdots$
  iii) $1 + x + x^2 + x^3 + \cdots$

b) $h(x) = \dfrac{2}{1+x} + \dfrac{3}{(1+x)^2} - \dfrac{1}{1-x}$

c) $h(x) = 4 - 9x + 10x^2 - 15x^3 + \cdots$

5 $g(x) = \dfrac{1}{(1-2x)^2} + \dfrac{2}{1-2x} + \dfrac{2}{1+2x}$

$5 + 4x + 28x^2 + 32x^3 + 144x^4 + \cdots$
$$-\tfrac{1}{2} < x < \tfrac{1}{2}$$

6 $f(x) = \dfrac{5}{1+2x} + \dfrac{4}{2-x} + \dfrac{3}{1-x}$

a) $\dfrac{5}{2}\ln|1+2x| - 4\ln|2-x|$
$$- 3\ln|1-x| + c$$

b) $10 - 6x + \dfrac{47}{2}x^2 - \dfrac{147}{4}x^3 \qquad -\tfrac{1}{2} < x < \tfrac{1}{2}$

CHAPTER 5
Revision Exercise

1 $\dfrac{x+10}{x(x+3)}$

2 $\dfrac{2}{x-2} + \dfrac{1}{x+1} \quad \ln 5$

3 a) $P = 3 \quad Q = 1 \quad R = -1 \quad S = 2$

b) $\dfrac{-107}{54}$

4 $\dfrac{3x-7}{(x+1)(x+2)} = \dfrac{13}{x+2} - \dfrac{10}{x+1}$

$13\ln 6 + 10\ln 4 - 23\ln 5$

5 i) $y = \dfrac{2}{2-x} + \dfrac{3}{(1-x)^2} - \dfrac{2}{1-x}$

ii) $6 + \dfrac{17}{2}x + \dfrac{45}{4}x^2 + \dfrac{113}{8}x^3 + \cdots$

6 a) $\dfrac{4}{x+4}$

b) $\dfrac{6x^2 + 13x - 5}{(x^2-4)(x+3)}$

c) $x^2 - 4 + \dfrac{4}{x^2+1}$

7 i) $\dfrac{1}{32}\left[\dfrac{1}{x-4} - \dfrac{2}{x} + \dfrac{1}{x+4}\right]$

ii) $\dfrac{1}{32}\ln\left(\dfrac{7}{135}\right)$

8 $2x + 3 + \dfrac{4}{2x+1} - \dfrac{3}{x}$

$x^2 + 3x + 2\ln|2x+1| - 3\ln|x| + c$

9 $f(x) = \dfrac{3}{1+x} - \dfrac{2}{1-x} + \dfrac{1}{1-2x}$

$2 - 3x + 5x^2 + 3x^3 + \cdots \qquad -\tfrac{1}{2} < x < \tfrac{1}{2}$

10 i) $\dfrac{1}{9} + \dfrac{1}{81}x^2 + \dfrac{1}{729}x^4 + \dfrac{1}{6561}x^6 + \cdots$

ii) $\dfrac{4x}{9-x^2}$

iii) $\dfrac{4}{9}x + \dfrac{4}{81}x^3 + \dfrac{4}{729}x^5 + \dfrac{4}{6561}x^7 + \cdots$

11 quotient $= 3x - 1$ remainder $= 4x - 7$.

12 $\lambda = 8, \ \mu = -9$

CHAPTER 6
Exercise 1

1 $x \sin x + \cos x + c$

2 $\dfrac{1}{5}xe^{5x} - \dfrac{1}{25}e^{5x} + c$

3 $\dfrac{1}{3}x \sin 3x + \dfrac{1}{9}\cos 3x + c$

4 $\dfrac{1}{5}x^5 \ln x - \dfrac{1}{25}x^5 + c$

5 $-x\cos x + \sin x + c$

6 $(2x+1)e^x + c$

7 $x^2 \sin x + 2x \cos x - 2\sin x + c$

8 $-\dfrac{1}{2}x^2 \cos 2x + \dfrac{1}{2}x \sin 2x + \dfrac{1}{4}\cos 2x + c$

9 $x \tan x + \ln|\cos x| + c$

10 $\dfrac{8}{3}\ln 2 - \dfrac{7}{9}$

11 $\dfrac{5}{27}e^3 - \dfrac{2}{27}$

12 $\dfrac{3}{4}e^{-2} - \dfrac{5}{4}e^{-4}$

13 $x \ln x - x + c$

14 SPs $(0,0)$ min $\quad (2, 4e^{-2})$ max
Area $= -10e^{-2} + 2$

16 $\pi(\pi^2 - 4)$

17 $\frac{3}{34}e^{3x}\sin 5x - \frac{5}{34}e^{3x}\cos 5x + c$

CHAPTER 6
Exercise 2

1 a) $\frac{1}{24}(4x - 3)^6 + c$

   b) i) $\frac{13}{3}$ **ii)** $-(2x - 1)^{-3} + c$

2 a) $\frac{1}{3}(x^2 + 4)^6 + c$

   b) i) $4\ln(x^2 + 25) + c$ **ii)** $\frac{98}{3}$

     iii) $\frac{1}{14}(x^2 + 4)^7 - \frac{1}{3}(x^2 + 4)^6 + c$

3 a) $\frac{-1}{6}(x^3 + 2)^{-2} + c$

   b) i) $4\sqrt{x^3 + 4} + c$

     ii) $\frac{1}{55}(x^5 + 4)^{11} + c$

     iii) $\frac{37}{42}$

     iv) $\frac{1}{60}(x^5 + 4)^{12} - \frac{4}{55}(x^5 + 4)^{11} + c$

4 a) $\sqrt{1 + e^{2x}} + c$

   b) i) $\frac{1}{20}(1 + 5e^x)^4 + c$

     ii) $\frac{1}{2}(4 + e^{-2x})^{-1} + c$

5 a) $\frac{1}{6}$

   b) i) $-\frac{1}{5}\cos^5 x + c$

     ii) $\frac{1}{16}\tan^8 2x + c$

6 b) $\frac{1}{24}$ **c)** $\frac{2}{99}$

7 a) $\frac{9}{800}$ **b)** 78 **c)** $\frac{45}{8}$

   d) $\frac{1}{4}$ **e)** $\frac{1}{4}$ **f)** $\frac{14}{3}$

CHAPTER 6
Exercise 3

1 $\frac{2367}{14}$ **2** $\frac{2}{35}$ **3** $\frac{9\pi}{4}$

4 $\ln 2 + \frac{11}{8}$ **5** $\frac{2 - \sqrt{2}}{3}$ **6** 5

7 $\frac{2}{3}$ **8** $\frac{2}{15}$

9 a) $\frac{\pi}{16}$ **b)** 16π

10 i) $2 - \sqrt{3}$ **ii)** $\frac{\pi}{3} - \frac{\sqrt{3}}{2}$

11 $\sin^{-1} x + c$

   $\frac{d}{dx}(\sin^{-1} x) = \frac{1}{\sqrt{1 - x^2}}$

12 a) $\tan^{-1} x + c$ **b)** $\frac{1}{1 + x^2}$

   c) $x\tan^{-1} x + \frac{1}{2}\ln(1 + x^2) + c$

CHAPTER 6
Exercise 4

1 $\frac{1}{2}x^2 + \frac{1}{4}x^4 + c$

2 $\frac{1}{4}xe^{4x} - \frac{1}{16}e^{4x} + c$

3 $\frac{2}{15}$

4 $\frac{1}{4}e^{4x} - e^{2x} + x + c$

5 $3\sqrt{2} + \sin x + c$

6 $\ln\left(\frac{12}{7}\right)$

7 $\frac{1}{2}\ln 2$

8 $\frac{506}{15}$

9 $\frac{15}{8} - 2\ln 2$

10 $\frac{16}{105}$

11 $\pi - 2$

12 $4(\sqrt{7} - \sqrt{3})$

13 $x^2 - x + \ln(x + 1) + c$

14 $\frac{32}{5}\ln 5 - \frac{31}{5}$

15 $\frac{61}{3}$

16 $\frac{1}{3}$

17 $\frac{1}{2}x^2 - 2x + 4\ln|x + 2| + c$

18 $4 - 2e^{\frac{1}{2}}$

19 $\frac{1}{2}\ln(x^4 + 1) + c$

20 $\frac{-2}{x + 2} + 3\ln\left|\frac{x - 2}{x + 1}\right| + c$

21 π

22 0.1585 (4 s.f.)

23 $x \ln(2x) - x + c$

CHAPTER 6
Revision Exercise

1 $4 - 6e^{-1}$

2 π^2

3 $\frac{-1}{15}(1 + 3 \cos x)^5 + c$

4 $x \tan x + \ln|\cos x| + c$

5 2

6 $\frac{1}{9}$

7 $\ln 3 + \frac{2}{3}$

8 $-\frac{3}{16}$

9 $\frac{2}{x} - \frac{1}{x-2}$

10 ii) $a^2\left(\frac{\pi}{3} - \frac{\sqrt{3}}{4}\right)$

12 ii) $\frac{58}{15}$

13 $\frac{1}{1+x} + \frac{8}{x-2} + \frac{12}{(x-2)^2}$
$6 - 7 \ln 2$

CHAPTER 7
Exercise 1

1 $y^2 = 4x$ **2** $y = \frac{4}{x}$

4 a) $x^2 + y^2 = 1$

b) $(x-3)^2 + (y-5)^2 = 1$

c) $\frac{x^2}{4} + y^2 = 1$

d) $\frac{x^2}{25} + \frac{y^2}{4} = 1$

e) $x^2 + y^2 = 25$

f) $\frac{(x-2)^2}{9} + \frac{(y-4)^2}{4} = 1$

$C_1: x = 4 \cos \theta, \ y = 2 \sin \theta$
$C_2: x = 3 + 2 \cos \theta, \ y = -4 + 2 \sin \theta$

7 gradient n intercept $(0, \ln k)$

10 a) $C(\theta, 1)$ **b)** $\theta - \sin \theta$

11 $\theta = -\frac{\pi}{3}, 0, \frac{\pi}{3}$ $(0, 2 - 2\sqrt{3})$
$(0, 2)$ $(0, 2 + 2\sqrt{3})$
$\theta = -\frac{\pi}{6}$ $(-5, 0)$

12 $A\left(0, \sqrt{\frac{\pi}{2}}\right)$ $C\left(0, -\sqrt{\frac{3\pi}{2}}\right)$
$B(-\sqrt{\pi}, 0)$ $D(\sqrt{2\pi}, 0)$
$(0, 0)$ $\left(\sqrt{\frac{\pi}{8}}, \sqrt{\frac{\pi}{8}}\right)$ $\left(-\sqrt{\frac{5\pi}{8}}, -\sqrt{\frac{5\pi}{8}}\right)$

CHAPTER 7
Exercise 2

1 $y = \frac{1}{3}x + \frac{1}{9}$ $y = 33 - 3x$

2 $y = 4x - 45$

3 $\frac{dy}{dx} = \frac{2t-1}{3t^2}$ $\left(\frac{1}{8}, -\frac{1}{4}\right)$

4 Tangent $y = -\frac{1}{4}x + 12$
Normal $y = 4x - 90$
$B\left(\frac{-3}{2}, 96\right)$

5 $y = -2x + 3$ $\left(-\frac{1}{2}, 4\right)$

6 $y = -px + p^3 + 2p$

7 $(e^{-1}, -e^{-1})$

8 $\frac{dy}{dx} = \frac{-2 \sin 2\theta - 2 \sin \theta}{2 \cos 2\theta + 2 \cos \theta}$
$(0, 3)$ $\left(\frac{\sqrt{3}}{2}, -\frac{1}{2}\right)$ $\left(-\frac{\sqrt{3}}{2}, -\frac{1}{2}\right)$
$\left(\frac{3\sqrt{3}}{2}, \frac{1}{2}\right)$ $\left(-\frac{3\sqrt{3}}{2}, \frac{1}{2}\right)$ $(0, 1)$

CHAPTER 7
Exercise 3

1 $4y^3 \frac{dy}{dx}$

2 $3e^{3y}\dfrac{dy}{dx}$

3 $3\cos 3x\dfrac{dx}{dt}$

4 $-3\sin 3\theta\dfrac{d\theta}{dr}$

5 $\dfrac{1}{y}\dfrac{dy}{dx}$

6 $2x^4y\dfrac{dy}{dx}+4x^3y^2$

7 $3x^2y^5+5x^2y\dfrac{dy}{dx}$

8 $2y\ln x\dfrac{dy}{dx}+\dfrac{y^2}{x}$

9 $3x^2e^{4y}+4x^3e^{4y}\dfrac{dy}{dx}$

10 $\dfrac{y^2+2y}{(y+1)^2}\dfrac{dy}{dx}$

11 $\dfrac{dr}{d\theta}\cos\theta-r\sin\theta$

12 $3\cos 3xe^{2y}+2\sin 3xe^{2y}\dfrac{dy}{dx}$

CHAPTER 7
Exercise 4

1 **a)** $\dfrac{dy}{dx}=\dfrac{x^2}{y^2}$

   b) $\dfrac{dy}{dx}=\dfrac{1}{y^2}$

   c) $\dfrac{dy}{dx}=\dfrac{4x-y+3}{x-2y}$

   d) $\dfrac{dy}{dx}=\dfrac{-y^3}{2x^2}$

   e) $\dfrac{dy}{dx}=\dfrac{1-\cos x}{1+\cos y}$

   f) $\dfrac{dy}{dx}=\dfrac{3x^2y^2-1}{2+2x^3}$

2 Tangent $y=-\dfrac{10x}{9}+\dfrac{32}{9}$

   Normal $y=\dfrac{9x}{10}-\dfrac{65}{10}$

3 **a)** $\dfrac{1}{(1+x)^2}$

   b) Tangent $y=-5x+9$

      Normal $y=\dfrac{1}{5}x+\dfrac{19}{5}$

4 $y=\dfrac{a-5}{a}x+25-5a$

   OP + OQ = 25

5 (2, 2)

6 (8, 2) (−8, −2)

8 At (1, 0) if $0<k<1$ tangent is
              horizontal
              if $k>1$ tangent is vertical
   At (0, 1) if $0<k<1$ tangent is vertical
              if $k>1$ tangent is horizontal

CHAPTER 7
Revision Exercise

1 **a)** $\dfrac{dy}{dx}=\dfrac{2t+2}{3t^2}$ **b)** $y=-\dfrac{1}{6}x+\dfrac{4}{3}$

2 **i)** $-\dfrac{4x+y}{x+2y}$ **ii)** (1, −4) or (−1, 4)

3 **a)** (−1, 0) (11, 0)   **b)** (0, −11) (0, 11)

   c) $\dfrac{dy}{dx}=\dfrac{3t^2-12}{2t}$ (3, −16) or (3, 16)

4 **ii)** R(−12, 1)   **iii)** $t=\dfrac{1}{2}$

5 −8

6 **i)** $t=2$ **ii)** 3

7 $-\dfrac{30}{59}$

8 **b)** (1, 3) and (3, 1)

9 $\dfrac{18-2\sqrt{27}\sin\theta}{3\sin^2\theta}$ $P\left(\dfrac{3\sqrt{3}}{8},\dfrac{39\sqrt{3}}{4}\right)$

10 **a)** $A\left(\dfrac{\pi}{2},0\right)$, $B\left(-\dfrac{3\pi}{2},0\right)$

   b) $t=0,\pi,2\pi$

11 $-\dfrac{1}{2}$ $\left(-\dfrac{8\sqrt{6}}{6},\dfrac{\sqrt{6}}{6}\right)$ $\left(\dfrac{8\sqrt{6}}{6},-\dfrac{\sqrt{6}}{6}\right)$

12 b) p^3 **c)** $Q\left(-p, \dfrac{4}{p^2}\right)$

CHAPTER 8
Exercise 1

1 $x = 2e^{2t} + 1$

2 $x = -3\cos 2t + 2\sin 2t + 8$

3 $x = te^t - e^t - 4$

4 $x = 3\ln(1 + \sin 2t) + 4$

5 $x = \ln(t + 1) + \dfrac{1}{2}t^2 - t + 3$

CHAPTER 8
Exercise 2

1 $y = \sqrt[3]{\dfrac{3}{2}x^2 - \dfrac{1}{2}}$

2 $y = e^{x^2}$

3 $y = \ln\left(\dfrac{1}{2}(x^2 + 1)\right)$

4 $y = \sqrt[3]{3x - 2}$

5 $y = 2e^{3x}$

6 $x = 20 + 5e^{0.2t}$

7 $y = \sqrt{72 - 2x^2}$

8 $y = x^2$

9 $y = 8e^{2x}$

10 $y = \left(x^2 - \cos 2x + 1\right)^2$

CHAPTER 8
Exercise 3

1 $y = \sqrt{x^2 + 1}$

2 $y = 4x + 3$

3 $y = \dfrac{-2}{\ln x + c} + 1$

5 $\cos y = A(1 - e^x)$

6 $\dfrac{1}{(x - 2)(8 - x)} = \dfrac{1}{6}\left[\dfrac{1}{x - 2} + \dfrac{1}{8 - x}\right]$

$y = \dfrac{4(x - 2)}{8 - x}$

7 a) $\sqrt{x^2 - 9} + c$ **b)** $y = x^2 - 9$

8 a) $\frac{1}{4}\tan 4x + c$ **b)** $y = \dfrac{1}{2\tan^4 x - 1}$

9 a) $-\dfrac{1}{3}xe^{-3x} - \dfrac{1}{9}e^{-3x} + c$

b) $\tan y = \dfrac{1}{9}(10 - 3xe^{-3x} - 1)$

CHAPTER 8
Exercise 4

1 c) $k = \ln\left(\dfrac{10}{9}\right) \approx 0.1054$

d) $x = 53.1$ **e)** 28.4 hrs

2 i) $x = 2t$ 4:00 p.m.

ii) $\dfrac{dx}{dt} = \dfrac{k}{x}$ 6:00 p.m.

iii) constant temperature throughout

3 i) b) $\theta = 12 + 86e^{-kt}$

$k = \dfrac{1}{10}\ln\left(\dfrac{86}{46}\right) \approx 0.06257$

d) 25 mins

ii) 1938

4 i) $\alpha = \dfrac{1}{10}, \beta = 1500$

ii) $y = 1500(1 - e^{-0.1t})$

iii) 16.1 weeks

5 ii) $N = 2050e^{kt}$ $k = \ln\left(\dfrac{3}{2}\right) \approx 0.4055 \ldots$

iii) $N = 470 + 1580e^{0.5t}$

iv) Second model grows much more quickly in long term

6 a) $\dfrac{dx}{dt} = -k\sqrt{x}$

c) 30 yrs

7 b) $\dfrac{1}{100}\left[\dfrac{1}{A} + \dfrac{1}{100 - A}\right]$

$\dfrac{1}{100}\left(\ln A - \ln(100 - A)\right) + c$

c) $p = \dfrac{1}{150}\ln\left(\dfrac{245}{95}\right) \approx 0.0063158$

d) 32.4

8 a) 257   **b)** 291

d) Model 2 gives a seasonal variation in population

9 iii) $3m^{\frac{1}{3}} = kt + c$ 10.78

10 b) $t = \dfrac{5}{3} \ln\left(\dfrac{33(p+20)}{7(100-p)}\right)$

i) 7.2

CHAPTER 8
Revision Exercise

1 $y = (\ln(4 + x^2))^2$

2 i) $\dfrac{dm}{dt} = -km$

iii) $k = -\dfrac{1}{30} \ln(\tfrac{85}{90}) \approx +0.001905$

iv) 171 (days)

As $t \to \infty$, $m \to 0$

3 $t > 36.06$

4 $y = \sqrt[3]{6x^2 - x^3 + 11}$

6 i) $\dfrac{dh}{dt} = -k\sqrt{h}.$ $2\sqrt{h} = -kt + c$

ii) 114.8

7 a) $\dfrac{dN}{dt} = kN$ **c)** 18.6

8 $\dfrac{1}{2}y^2 + y = x - \dfrac{1}{2}x^2 + c$

9 $\dfrac{dC}{dt} = -kC$

i) £567   **ii)** 23

10 $y^3 = x^3 + \dfrac{3}{x} - \dfrac{3}{2}$

REVISION 1
Algebra and Graphs

1 $1 - 4x + 12x^2 - 32x^3 + \cdots$

$-\dfrac{1}{2} < x < \dfrac{1}{2}$

2 $2 - \dfrac{3}{4}x - \dfrac{9}{64}x^2 + \cdots$

$-\dfrac{4}{3} < x < \dfrac{4}{3}$

3 a) $\dfrac{x(x-8)}{2(x-2)}$ **b)** $\dfrac{6x^2 - 11x - 1}{(x+2)(x-1)^2}$

4 $1 - x - x^2 + \cdots$

5 a) i) $(1 + 2x)^{-1} = 1 - 2x + 4x^2 - 8x^3 + \cdots$

$-\dfrac{1}{2} < x < \dfrac{1}{2}$

ii) $(1 - 5x)^{-1} =$

$1 + 5x + 25x^2 + 125x^3 + \cdots$

$-\dfrac{1}{5} < x < \dfrac{1}{5}$

b) $\dfrac{3}{1 + 2x} + \dfrac{2}{1 - 5x}$

c) $5 + 4x + 62x^2 + 226x^3 + \cdots$

$-\dfrac{1}{5} < x < \dfrac{1}{5}$

6 $(0, 4 - e^3)$ $(0, 4 - e^{-3})$

$((\ln 4)^2 - 9, 0)$

7 a) $1 + x - \dfrac{1}{2}x + \cdots$

b) $1 + 2x + 6x^2 + \cdots$

$1 + 3x + \dfrac{15}{2}x^2 + \cdots$

$\sqrt{17} \approx 4.123$

8 $\dfrac{1}{2}\left[\dfrac{3}{x} - \dfrac{2}{x+1} + \dfrac{1}{x+2}\right]$

$\dfrac{1}{2}\ln(\tfrac{128}{27})$

9 a) $\dfrac{2}{x+1} + \dfrac{1}{x+2} - \dfrac{1}{(x+2)^2}$

10 a) $a = 8$ **b)** $\alpha = \dfrac{3}{2}, -\dfrac{1}{2}$

12 $3x^2 - 4x + 5 + \dfrac{3}{x+2}$

$56 - 3 \ln 2$

13 i) $1 + 2x - 2x^2 + \cdots$

ii) $-\dfrac{1}{4} < x < \dfrac{1}{4}$

iii) $k = 5$ 8

14 $\dfrac{1}{3} + \dfrac{2}{27}x + \dfrac{2}{81}x^2 + \cdots$

$-\dfrac{9}{4} < x < \dfrac{9}{4}$

REVISION 2
Differentiation and Integration

1 i) $\dfrac{dy}{dx} = \dfrac{2t + 3t^2}{1 + 2t}$

ii) $5y - 16x + 36 = 0$

2 $12y - 5x + 22 = 0$

3 $\left(2, \dfrac{2}{e}\right)$ max

4 **a)** $9 \cos 3x (1 + \sin 3x)^2$

   b) $\dfrac{-8 \sin 4x}{3 + 2 \cos 4x}$

5 $\sin x - \dfrac{2}{3} \sin^3 x + c$

6 $6x \cos(x^3) - 9x^4 \sin(x^3)$

7 **i)** $\dfrac{1}{6}x^6 \ln x - \dfrac{1}{36}x^6 + c$

   ii) $\dfrac{1}{480}$

8 $y = \dfrac{\sqrt{3}}{12}x + 1 - \dfrac{\sqrt{3}\pi}{72}$

10 **a)** $y = \dfrac{1}{16}x^2$ **c)** $Q\left(-\dfrac{68}{3}, \dfrac{289}{9}\right)$

11 $-\dfrac{1}{2}$

12 **a)** $C\left(e^{-\frac{1}{2}}, -\dfrac{1}{2}e^{-1}\right)$ **c)** $\dfrac{2}{9}e^3 + \dfrac{1}{9}$

   d) $\dfrac{5e^3 - 9e - 2}{18}$

13 $\sqrt{2}\left(1 + \dfrac{\pi}{4}\right)$

14 **a)** $(\ln 17, 0)$, $\left(0, \dfrac{1}{2}\ln 17\right)$

   c) $\text{grad} = -\dfrac{1}{16}$

15 **a)** $-\pi$ **b)** $\dfrac{\pi}{4}$

16 $\dfrac{1}{6}(4 + 3 \ln x)^2 + c$

19 **iii)** $12x^2 - 3$ $\dfrac{-3 \sin 3\theta}{\sin \theta}$

20 **ii)** $\sqrt{3} - 1 - \dfrac{\pi}{12}$

21 $\dfrac{8}{3}\ln 2 - \dfrac{7}{9}$

22 **ii)** $\dfrac{1}{12}$

23 **i)** $A(0, 3)$ $B(3, 0)$

   ii) $C\left(5, -2e^{-\frac{5}{2}}\right)$

   iii) $4e^{-\frac{3}{2}} + 2$

24 **i)** $A(0, 2)$ $B\left(\dfrac{\pi}{6}, 0\right)$

26 **i)** $x \ln(3x) - x + c$ **ii)** $\dfrac{\pi}{12}$

27 **i)** $-x \cos x + \sin x + c$ **ii)** $\dfrac{\sqrt{3}}{3}$

29 **a)** $y = 4x + 1 - \dfrac{\pi}{2}$

   b) $\dfrac{1}{4}\ln 2$

30 **ii)** **a)** $\tan \alpha = \dfrac{3}{4}$

     b) $\theta = -0.232$ or 2.086

31 $\ln(\ln x) + c$

32 Normal $y = -\dfrac{1}{p}x + 3p + 2p^3$

   $Q\left(\dfrac{3p^2}{4}, -\dfrac{p^3}{4}\right)$ $P = \pm\sqrt{2}$

REVISION 3
Differential Equations

1 **i)** $-x^2 \cos x + 2x \sin x + 2 \cos x + c$

   ii) $\tan^2 x$

   iii) $\tan y - y =$

      $-x^2 \cos x + 2x \sin x + 2 \cos x + c$

2 $\dfrac{dr}{dt} = \dfrac{k}{r^2}$ 2.96 s

3 **i)** $T = -6x + 470$ $360°$

   ii) $T = 380 - \dfrac{1}{10}x^2$

4 **ii)** $x = \dfrac{-1}{kt + c}$

   iii) As $t \to 2$, $x \to \pm\infty$

5 **a)** $t = 7.05$ s   **b)** $t = 14.03$ s

6 $n = 50 + 450e^{2t}$

   exponential growth not sustainable

7 **i)** $P = P_0 e^{\sin kt}$

   ii) $P_{max} : P_{min}$ is $e^2 : 1$

9 **i)** $\alpha = 15$ $\beta = 0.2$

   ii) $\theta = 95 - 75e^{-0.2t}$

     $t = 1.116$ mins

   iii) As $t \to \infty$, $\theta \to 95°$

10 **d)** $t = 20\,000 \ln\left(\dfrac{40\,000}{40\,000 - r^2}\right)$

    or $r = \sqrt{40\,000 - 40\,000\,e^{-\frac{t}{20\,000}}}$

    As $t \to \infty$, $r \to 200$

   e) 11.5 days

11 i) 340 s   **ii) b)** 378.5 s

12 $y = \ln\left(\dfrac{8}{9e^{-x} - 1}\right),\quad \ln\left(\dfrac{8}{3}\right)$

REVISION 4
Vectors

1 b) $\sqrt{337}$ **c)** 14

2 a) $\underline{r} = \begin{pmatrix} 6 \\ 3 \\ 0 \end{pmatrix} + t\begin{pmatrix} 2 \\ 4 \\ 4 \end{pmatrix}$

b) $(5, 1, -2)$ **c)** $22.5°$

3 a) $\begin{pmatrix} 2 \\ 1 \\ -2 \end{pmatrix}$ **b)** 3   **d)** $\lambda = -3$

4 i) $\underline{r}_2 = \begin{pmatrix} 0 \\ 8 \\ 0 \end{pmatrix} + s\begin{pmatrix} -16 \\ -8 \\ 4 \end{pmatrix}$

ii) $P(0, 4, 4)$

iii) $\underline{r}_2 = \begin{pmatrix} 16 \\ 8 \\ 0 \end{pmatrix} + t\begin{pmatrix} -16 \\ -4 \\ 4 \end{pmatrix}$

v) 48   **vi)** $62.4°$

5 i) $71.6°$

6 i) $a = 4$ $b = -3$
iii) $(12, -2, -1)$ or $(0, 14, -1)$

7 ii) $\begin{pmatrix} -2 \\ 1 \\ -9 \end{pmatrix}$ $\sqrt{86}$

iii) $\underline{r} = \begin{pmatrix} 4 \\ -3 \\ 2 \end{pmatrix} + s\begin{pmatrix} -2 \\ +1 \\ -9 \end{pmatrix}$

$(6, -4, 11)$

8 i) $67.6°$ **ii)** $(5, 0, -1)$
9 i) $32.3°$ **ii)** $(1, -4, 14)$

Sample exam paper

1 $\pm \dfrac{1}{\sqrt{6}}(\underline{i} - \underline{j} + 2\underline{k})$

2 $x = \dfrac{5\pi}{12}$

3 i) $\dfrac{4 - x}{2 - x}$ **ii)** $2 + \dfrac{1}{2}x + \dfrac{1}{4}x^2$

4 $e - 1 + 3\ln\left(\dfrac{2}{e + 1}\right)$

5 i) Quotient $x^2 + 3x - 2$ remainder -5

ii) $2x + 3 + \dfrac{10}{(x + 1)^3}$

6 a) $\tan x + 2\ln|\sec x| + c$

b) $\dfrac{x^5}{25}(5\ln x - 1) + c$

7 i) $(4, 9, 5)$ **ii)** $53.3°$

8 $akt = \ln\left|\dfrac{2a - x}{2(a - x)}\right|$

9 ii) $\left(-13\dfrac{14}{27}, 13\dfrac{13}{27}\right)$ **iii)** $x^2 - y^2 = 1$

INDEX